SOCIOECONOMIC ASPECTS OF RENEWABLE ENERGY TECHNOLOGIES

The World Employment Programme (WEP) was launched by the International Labour Organisation in 1969, as the ILO's main contribution to the International Development Strategy for the Second United Nations Development Decade.

The means of action adopted by the WEP have included the following:
— short-term high-level advisory missions;
— longer-term national or regional employment teams; and
— a wide-ranging research programme.

Through these activities the ILO has been able to help national decision-makers to reshape their policies and plans with the aim of eradicating mass poverty and unemployment.

A landmark in the development of the WEP was the World Employment Conference of 1976, which proclaimed inter alia that "strategies and national development plans should include as a priority objective the promotion of employment and the satisfaction of the basic needs of each country's population". The Declaration of Principles and Programme of Action adopted by the Conference will remain the cornerstone of WEP technical assistance and research activities during the 1980s.

This publication is the outcome of a WEP project.

SOCIOECONOMIC ASPECTS OF RENEWABLE ENERGY TECHNOLOGIES

EDITED BY
Ramesh Bhatia AND Armand Pereira
IN COLLABORATION WITH
MADHURA SWAMINATHAN

A STUDY PREPARED FOR THE
INTERNATIONAL LABOUR OFFICE WITHIN THE
FRAMEWORK OF THE
WORLD EMPLOYMENT PROGRAMME

PRAEGER

New York
Westport, Connecticut
London

The designations employed in ILO publications, which are in conformity with United Nations practice, and the presentation of material therein do not imply the expression of any opinion whatsoever on the part of the International Labour Office concerning the legal status of any country, area, or territory or of its authorities, or concerning the delimination of its frontiers.

The responsibility for opinions expressed in studies and other contributions rests solely with their authors, and publication does not constitute an endorsement by the International Labour Office of the opinions expressed in them.

Reference to names of firms and commercial products and processes does not imply their endorsement by the International Labour Office, and any failure to mention a particular firm, commercial product, or process is not a sign of disapproval.

Library of Congress Cataloging-in-Publication Data

Socioeconomic aspects of renewable energy technologies: a study prepared for the International Labour Office within the framework of the World Employment Programme/edited by Ramesh Bhatia and Armand Pereira in collaboration with Madhura Swaminathan.

p. cm.
Bibliography: p.
Includes index.
ISBN 0-275-92851-9 (alk. paper)
1. Renewable energy sources—Developing countries—Case studies.
2. Alcohol fuel industry—Developing countries—Case studies.
3. Irrigation pumps—Developing countries—Case studies.
I. Bhatia, Ramesh, 1944- . II. Pereira, A. (Armand)
III. Swaminathan, Madhura. IV. International Labour Organisation.
V. World Employment Programme.
HD9502.D442S63 1988
333.7'09172'4—dc19 87-22890

Library of Congress Catalog Card Number: 87-22890
ISBN: 0-275-92851-9

First published in 1988

Praeger Publishers, One Madison Avenue, New York, NY 10010
A division of Greenwood Press, Inc.

Printed in the United States of America.

The paper used in this book complies with the Permanent Paper Standard issued by the National Information Standards Organization (Z39.48-1984).

10 9 8 7 6 5 4 3 2 1

Contents

List of Tables

Abbreviations and Acronyms

CURRENCIES

$	United States dollar
Cr$	Brazilian cruzeiro
Crore	10 million Rs, India
Escudo	Cape Verde
Ks	Kenyan shilling
Lakh	100 thousand Rs, India
LE	Egyptian pound
Paise	1/100 Rs, India
Rs	Rupee (India and Sri Lanka)
S£	Sudanese pound
Taka	Bangladesh

UNITS OF MEASUREMENT

bbl	barrel
cm (cm^2, cm^3)	centimeter (square, cubic)
ft^3	cubic feet

ha	hectare
hp	horsepower
hp-hr	horsepower-hour
kg	kilogram
km	kilometer
l	liter
lat	latitude
lb	pound
lpd	liters per day
l/t	liters per ton
l/yr	liters per year
LPG	liquid petroleum gas
m (m^2, m^3)	meter (square, cubic)
m.bbl	million barrels
Mcal	mega calories
m.ha	million hectares
m.l	million liters
m.t	million tons
min	minute
mm	millimeter
rpm	rotations per minute
sec	second
t	ton, metric ton
toe, ktoe	tons of oil equivalent, thousand toe
t/ha	tons per hectare
V, kV	volt, kilovolt
W, mW, kW, MW, TW	watt, milliwatt, kilowatt, megawatt, terawatt, respectively
W_p, kW_p	peak watt, peak kilowatt
kWh	kilowatt-hour

OTHER ABBREVIATIONS AND SYMBOLS

AC	alternating current
B/C	benefit/cost (ratio)
bhp	break horsepower
c.i.f	cost, insurance, freight
CU	capacity utilization

DC	direct current
E_a	field application efficiency (of water in irrigation)
EHH	electrified households
EIRR	economic internal rate of return
E_{crop}	evapo-transpiration rate of particular crop
ET_o	evapo-transpiration rate
ER	effective rainfall
f.o.b	free on board
hr (hrs)	hour (hours)
hrs/yr	hours per year
IRR	internal rate of return
Iw	irrigation water
K_c	crop coefficient
MP-1	type of windmill developed by the National Aeronautics Laboratory, India
MPPT	maximum power point tracker
NEHH	non-electrified households
NPV	net present value
OMA	operation, maintenance, and administration
o.m.r.l	operation, maintenance, repair, and lubrication
PV	present value
R and D	research and development
SRD	social rate of discount
TDN	total daily nutrient
t/ha	tons per hectare
THE-II	type of windmill developed by CWD (see below)
12-PU500	type of windmill known also as the Ghazipur or the Ghazipur/ Allahabad windmill, India
WEU I/2	type of windmill developed by the Wind Energy Utilisation Project, Sri Lanka
whp	water horsepower
WP-2	type of windmill developed by the National Aeronautics Laboratory, India
"—"	symbol used in tables meaning that data are either not applicable or not available
"*"	symbol used in tables to refer to a single note that is applicable to a whole table

ACRONYMS

AFPRO	Action for Food Production
ARDC	Agricultural Refinance and Development Corporation (Bombay, India)
BADC	Bangladesh Agricultural Development Corporation
BIS	Birla Institute of Technology and Science (Pilani, India)
CEL	Central Electronics Limited (New Delhi, India)
CENAL	Executive Committee for National Alcohol (Brazil)
CNPq	National Research Council (Brazil)
C/S	Center-South (a macro-region in Brazil)
COPPERSUCAR	Cooperative of Sugar and Alcohol Producers (State of Sao Paulo, Brazil)
CPRI	Central Power Research Institute (Bangalore, India)
CSIR	Council of Scientific and Industrial Research (India)
CWD	Consultancy Services Wind Energy Developing Countries (Amersfoort, The Netherlands)
ESCAP	Economic and Social Commission for Asia and the Pacific
FAO	Food and Agriculture Organization (Rome)
GAU	Gujarat Agricultural University (India)
HP	(State of) Himachal Pradesh, India
KRECS	Kotputali Rural Electric Cooperative Society (Rajasthan, India)
KVIC	Khadi and Village Industries Commission (Bombay, India)
NAL	National Aeronautics Limited (Bangalore, India)
IAA	Institute of Sugar and Alcohol (Brazil)
IARI	Indian Agricultural Research Institute
ICAR	Indian Council of Agricultural Research (New Delhi, India)
ILO	International Labor Organization (Geneva, Switzerland)
ISA	International Sugar Agreement
MIC	Ministry of Industry and Commerce (various countries)
MME	Ministry of Mines and Energy (various countries)
N/NE	North/North-East (a macro-region in Brazil)
NE	North-East (a region within the N/NE macro-region)
PETROBRAS	Brazil's national oil company
PLANALSUCAR	IAA's Sugar Cane Research Center (Brazil)
PRAD	Planning Research and Action Division (India)
RTI	Rural Technology Institute (Allahabad, India)
SDCA	Swiss Development Cooperation Agency (Berne, Switzerland)
SEI	Solar Electric International

SWRC	Social Work Research Center (India)
TOOL-ORP	TOOL Foundation–Organization of Rural Poor (Allahabad, India)
UN	United Nations (New York, Geneva)
UNCTAD	United Nations Conference on Trade and Development (New York, Geneva)
UNDP	United Nations Development Program (New York)
UNIDO	United Nations Industrial Development Organization (Vienna, Austria)

Preface

Significant changes have occurred in both the composition of energy supply and the choice of technology, particularly in activities where energy intensity has an important bearing on the costs and competitiveness of goods and services. In the context of these changes, the present volume, edited and partly written by Armand Pereira and Ramesh Bhatia, in collaboration with Madhura Swaminathan, includes case studies on the macroeconomic evaluation of ethyl-alcohol programs and the microeconomic evaluation of technologies for irrigation pumping.

Shifts away from oil to more labor-intensive fuels, notably coal and biomass energy sources, have had a significant effect on employment. For example, among the renewable energy sources with large potential for direct substitution of oil products, ethyl alcohol from biomass is much more labor intensive than any major conventional energy source, including coal, petroleum, and natural gas. Also, the increased exploration for and production of oil and natural gas in non-OPEC countries has led to a greater geographical diversification of world fuel supplies and related employment. The direct employment impact of this capital-intensive activity is not large, but it has significant effects on manpower training, on institutional and technological capacity building, and, in some countries, on the development of local capital goods sectors.

However, changes in technology outside the energy sector, which have been partly induced by energy price increases, have not had a very significant direct impact on employment, although some ILO country studies on energy price effects have revealed, for example, that after the two oil price shocks of 1973–74 and 1978–80 there were some shifts from mechanized to manual techniques in agriculture as well as substitutions of biomass for oil-derived fuels in industry. It appears, however, that such a substitution of labor for energy was rare and temporary, not only because of relative factor prices but also because, in many cases, the labor-intensive techniques do not perfectly replace the more energy- and/or capital-intensive ones.

Important issues concerning the choice of energy-intensive technology arise in connection with the indirect employment and income effects of alternative resource allocation. In irrigation pumping, for example, the optimum technology for one set of end-use conditions will not be optimum for another (due to variations in farm size, cropping pattern, water and energy supply, repair and maintenance capacity, opportunity costs of labor and other inputs, etc.). Government policies that guide farmers' decisions on irrigation technology choice, however, are often too standard and fail to reflect the varying attractiveness of the options available for different conditions.

Another important issue is the development of management training programs to improve decision-making capacity for the evaluation and selection of technologies for various end-uses and -user conditions. Organizational and institutional aspects of the development and diffusion of renewable energy technologies are equally important. While these aspects are beyond the scope of this book, it is hoped that the analyses of socioeconomic aspects of the selected technologies and end-uses included will be useful for future work on training and other institutional issues.

The research on which this book is based was financed out of grants from the governments of the Netherlands and Sweden (Swedish Agency for Research and Economic Cooperation) to the ILO Technology and Employment Programme.

A. S. Bhalla,
Chief,
Technology and Employment Branch,
International Labour Organisation

Acknowledgments

Many institutions and individuals have contributed to this volume and it is impossible to name them all.

The editors are especially indebted, however, to Ajit S. Bhalla, who encouraged and assisted the preparation of the book from its conception to its publication, as well as Peter Duiker and Fred Fluitman, who commented on a previous draft.

Russell de Lucia (de Lucia and Associates Inc., Cambridge, Massachusetts) deserves a very special word of thanks for his useful suggestions on the technical coverage and structure of the volume.

The editors would also like to thank Michele Bhunnoo, Gaby Thevenon, Anne Meade and Susan Saidi for typing, and Frances Williams for her editing suggestions.

1

Introduction

Armand Pereira and Ramesh Bhatia

As world oil prices rise and fall, it is inevitable that energy policymakers and researchers shift their attentions and priorities. The continuing instability in the world oil market makes it necessary, however, to maintain a minimum base of information that can be updated and used as guidelines for planning. One of the consequences of the recent decline in oil prices is that some of that information base has become outdated. But another problem is that much of the overall project experience with renewable energy technologies over the 1978–1984 period, in particular, had not been evaluated by the time energy policy and research priorities began to shift as a result of the fall in world oil prices.

Renewable energy technologies for different applications have been field-tested, and some have been used commercially, in many developing countries since the mid-1970s. Also a large volume of literature has appeared on energy policy issues and options and several studies have been carried out on the socioeconomic aspects of renewable energy sources used for various purposes.

However, much of the literature has emphasized the supply of specific technologies without adequate analysis of users' needs and different options to meet them. Ex-post evaluations of recent experiences with renewable technologies are still scarce, and many of those that have appeared provide limited insights for planning and implementation in different country situations. Some methodological questions in ex-ante economic evaluations of renewable technologies have not yet been settled.

COVERAGE, OBJECTIVES, AND FOCUS

No single volume on renewable energy technologies can cover all technologies and all their relevant aspects of planning and implementation. This book is no exception, and therefore some choices had to be made about its coverage and focus.

The volume is aimed at energy policy-makers and covers a selected number of technologies and topics in the following categories.

- substitution of renewable energy sources for petroleum products in the transport sector;
- provision of decentralized energy for productive activities in rural and other isolated areas.

These issues are analyzed separately in Parts I and II of this volume, taking into account energy policymaking needs of developing countries and the gaps in the literature.

Part I covers ethanol production programs and focuses on *macroeconomic evaluation.* It includes an ex-post assessment of the impact of the Brazilian ethanol program in Chapter 2 and an ex-ante evaluation of the implications of expanding ethanol production in India in Chapter 3. Employment and income distribution effects are analyzed together with effects on petroleum substitution and balance of payments, which are the main socioeconomic aspects of ethanol programs.

Part II covers technology options for irrigation pumping. Human- and animal-powered pumping devices in Chapter 4, windmills in Chapter 5, and solar-thermal, solar-photovoltaic, and dual-fuel (diesel/biogas and diesel/producer-gas) pumping systems in Chapter 6 are compared against diesel/petrol and electric pumping technologies in a number of countries of Asia and Africa. In those chapters, the focus is on *microeconomic evaluation* of the alternative technologies.

The final chapter summarizes the conclusions of the individual chapters, including their implications for policy, further research, and training. The informal table below shows the outline of the volume.

	Category of issues	Main end-uses	Options considered	Focus	Chapters
PART I	Large-scale oil substitution	transport and chemicals	ethanol vs. oil products	macro-economic (ex-post plus ex-ante)	2 and 3

PART II	Energy for productive activities in rural areas	irrigation	oil products or electricity vs. wind, solar, plus dual-fuel	micro-economic (ex-post plus ex-ante)	4, 5, and 6

Neither the categories of issues nor the technology options selected under each category (see above) are necessarily separable. However, the patterns of demand for energy, and both the supply and costs of energy resources and related technologies in developing countries, make these two categories distinguishable in various ways. Since some recent studies have already mapped issues and options in a systematic manner,[1] the two sections below on each of the two parts will be limited to a general discussion of the topics addressed, including literature review and methodological considerations, as appropriate.

PART I. MACROECONOMIC EVALUATION OF ETHANOL PRODUCTION

Ethanol (biomass ethyl alcohol) remains the most suitable option for direct substitution of petroleum-derived fuels in the transport sector. It is both technologically mature and economically promising.[2] This is one reason why it is selected for analysis in this volume, but another reason is to rectify the gaps that presently exist in the literature.

Several studies (some of which are discussed below) have addressed different aspects of ethanol production, but in-depth assessments of specific country experiences are yet to appear in the international literature. Moreover, there are still some methodological questions regarding ex-ante evaluation of ethanol programs.

Anhydrous ethanol (which is practically water-free) can substitute for petrol in blends up to 20–22 percent (i.e., "gasohol") without requiring any mechanical changes in vehicles and without reducing fuel efficiency in kilometers-per-litre. Hydrous ethanol (4 percent water) can be consumed as straight fuel in vehicles with minor engine adaptation or in new vehicles equipped with Otto-cycle engines, although its efficiency is about 17–18 percent lower than that of petrol or gasohol. It is also technically feasible to add anhydrous ethanol to diesel oil in combustion engines in blends of about 5 percent,[3] but ethanol is 32–39 percent less efficient than diesel, whereas petrol is only about 9 percent less efficient than diesel. Thus, for creating diesel blends, petrol is cheaper than ethanol. Ethanol may nevertheless substitute indirectly for diesel through changes in the relative composition of diesel- and petrol/ethanol-powered vehicles. It may also substitute for diesel in irrigation pumping with minor adaptations. In addition, ethanol can also be used as a feedstock for human beverages, for pharmaceuticals, and for some major petrochemicals.

Ethanol can be competitive with petroleum products in many countries at the oil price levels of 1979–1981 and, in some countries, even at the oil price levels of 1982–1984.[4] Moreover, the substitution of ethanol for petroleum imports can have positive effects on the balance of payments and on economic growth. Its direct employment impact may be very significant where ethanol is produced from relatively labor-intensive crops (e.g., sugar cane and manioc), and crop substitutions may be minimal or even generate a net gain in employment. The spillover effects may be significant on the capital goods sector and on research and development (R & D) in particular. In addition, ethanol industries may provide a number of by-products that can contribute to rural development. Each liter of ethanol from sugar cane yields about 12 liters of effluent (commonly referred to as "stillage"). If thrown into rivers and lakes, stillage has a highly polluting effect, but it can be treated in closed reservoirs and be partly used for direct irrigation and fertilization. It can also be used to produce biogas for use in adjacent communities and rural industries. A by-product of sugar-cane production, bagasse, is used for heat generation, but there is usually a surplus, depending on the burning efficiency of the heat generation. Both surplus bagasse and stillage biogas can be used to generate electricity. Surplus bagasse can also be used as animal feed and/or as input to chipboard industries.

Such considerations have led a number of countries to initiate production of ethanol as fuel and/or material for the chemical industry. Besides Brazil, which has the largest commercial-scale fuel-ethanol program based on sugar cane and molasses, India has over 100 small distilleries using molasses, producing mainly for the chemical industry. Other countries with some production in 1982–1984 were Kenya, the Philippines, Zimbabwe, Sri Lanka, Australia, Thailand, Indonesia, the United States, and Malawi. Also Mali, Burkina Faso (Upper Volta) and Nicaragua initiated some limited production in 1985.[5] A number of other countries in Central America, the Caribbean, and Australasia had plans for production in 1984/85, although implementation will have been discouraged by the oil price fall in 1986. At this price level ($12–18 per barrel of crude oil), there is no ethanol production process for the replacement of fuels and petrochemicals that is likely to be economic, including the cheapest production process based on molasses.

Despite the general attractiveness of ethanol as a substitute for petroleum products (when oil prices make its economics favorable), not all developing countries have appropriate conditions for commercial-scale production. Even some of those that do may find it economically unattractive to produce ethanol on a large scale. Some other countries may choose to produce ethanol for strategic reasons, even if it is more expensive than oil imports, because of continuing instability in the oil market and for promoting self-reliance. However, in all countries that may have sound justifications for producing ethanol on a commercial scale, production is complicated by linkages between foreign trade, agriculture, oil refining, petrochemicals, fertilizers, distillery equipment and other basic inputs, fuel distribution, and so on. Planning requires explicit consideration of such linkages in developing countries, where rigidities in production structures tend to exist and market prices rarely reflect opportunity costs, due to imperfections of the market system and government price controls.

Some studies have tried to identify countries where ethanol would be relevant. For example, Rask (1979) developed a methodology, adopted by the World Bank (1980b) and FAO (1980), whereby countries were placed in one of four categories, ranging from one of energy-surplus/agricultural-surplus to one of energy-deficit/agricultural-deficit. Countries that were classified as energy-deficit/agricultural-surplus were presumably those where ethanol programs would be most relevant. This methodology has serious limitations. The concepts of agricultural and energy "surplus" and "deficit" are both irrelevant and misleading in the context of planning for ethanol, because of trade considerations and/or rigidities in both the relative supply of different fuels and in interfuel substitution. Countries may have a surplus of biomass or coal or natural gas, but they may have deficits of certain transport fuels. Some countries may have an "agricultural deficit" because land is badly managed. Others may have certain land areas that are unsuitable for cultivating food crops, but feasible for some ethanol feedstocks with low or no food value, such as wood. Ethanol produced from molasses does not require any land resources other than those required for sugar production. Hence there may be sugar-producing countries with agricultural and/or food deficits, but with surplus molasses at low opportunity cost to sustain a minimal level of production. By the same token, countries with a large manioc-starch industry, for example, may find it attractive (depending on market demand and opportunity cost) to develop an ethanol industry based on this feedstock.

A study by the World Bank (1980b) provided a useful analysis of economic parameters for investments in ethanol distilleries, based on certain assumptions of capital costs, crop yields, distillery-gate costs of raw materials and ex-refinery costs of petrol. The major conclusion of this World Bank study was that "the relative economic merits of alcohol production will vary among countries depending on the specific economic parameters of their agricultural, industry and energy sectors."

Koide et al. (1982) reported on ethanol production developments in South-East Asia, and analysed future prospects for the region. It is inconclusive about these prospects but provides a useful problem-finding survey and gives ideas for further investigation at the country level.

Although the above studies have focused on different aspects of ethanol production, in-depth socioeconomic analyses of ethanol programs have not appeared in the international literature.[6] Ex-post assessments are necessary to evaluate the impact of observed production and identify strategies and policies to improve this impact in the future. Prospective or ex-ante assessments must be used for country situations where the options of initiating or expanding ethanol production are being considered. The former assessments may require certain assumptions due to the lack of some data, but on the whole the analysis can be based on observed events. Ex-ante assessments, however, require an analysis of different scenarios, based on assumptions about the (a) supply, demand, and prices of oil products; (b) input requirements for producing ethanol (including technology, capital, land, labor, etc.); (c) opportunity costs of these inputs; and (d) organization in the supply of ethanol and inputs, which will determine the distribution of benefits from production. Since the conclusions of these types of assessment depend on the assumptions made, these must be explicitly stated and sensitivity analyses should be included.

An ex-post assessment of Brazil's ethanol program can contribute to a better understanding of the order of impact of commercial-scale programs and of problems inherent particularly in their implementation. No other ongoing ethanol program provides as many insights as that of Brazil. On the other hand, an ex-ante assessment of an expansion of ethanol production for fuel purposes in a country with scarce land resources such as India will bring out some of the major complexities in planning.

Some general remarks about the main differences between the two chapters in terms of coverage, emphasis, and methodology are in order. In Chapter 2 on Brazil, the impact in selected years is estimated whereas, in Chapter 3, the objective is to assess the social profitability of a new ethanol program. This implies some differences in relative emphasis on issues. In Chapter 2, balance of payments, employment, and income distribution are given detailed analysis since they are the key effects, whereas in Chapter 3 the emphasis is on overall social costs and benefits. Their emphasis also implies different methodologies and limitations. In Chapter 2, those effects are calculated on the basis of available statistics and field surveys. Chapter 3 is much less precise about these effects but provides a more global view of ethanol production.

In the ex-ante evaluation in Chapter 3, the social costs and benefits of the programs are expressed in local currency for comparison and estimation of net present values. This requires the use of a shadow exchange rate (i.e., official exchange rate plus a standard premium[7]) to reflect the existing scarcity of foreign exchange. The shadow exchange rate is used to convert into local currency the prices (in foreign exchange) of traded goods associated with the programs (i.e., fertilizer, fuels, and other inputs imported at the margin, as well as oil import savings and fuel exports due to ethanol production). In India, a standard premium of 25 percent on foreign exchange has been used by the government, and hence it is used in Chapter 3. In the ex-post analysis of the impact of ethanol production in Brazil, the focus requires a different method. Inputs imported at the margin and benefits in foreign exchange are expressed in U.S. dollars, thus not requiring the use of shadow exchange rate for this purpose. Nevertheless the study discusses capital costs (including domestic and foreign-exchange components) as well as costs of ethanol production, which are expressed in U.S. dollars by using the standard conversion factor.[8]

PART II. COMPARATIVE MICROECONOMIC EVALUATION OF RENEWABLE ENERGY TECHNOLOGIES FOR IRRIGATION PUMPING

Water pumping for irrigation has been selected because of (a) the importance of this activity for rural development, (b) the broad field experience with most of the technologies addressed, and, in particular, (c) the need for improving the methods of comparative economic evaluation.

In principle, oil and conventional sources of electricity can provide energy for productive activities and basic household needs in rural or other isolated areas but, in practice, this has been constrained by *financial*[9] and *economic*[10] factors. Financial constraints to users may be reduced by government subsidies, but economic costs of supplying oil and electricity to rural areas can be very high on account of dispersed activities and low load factors, which are determined by locational patterns, development levels, and availability of complementary resources. On the other hand, most energy options that can have a significant impact on oil substitution are likely to be developed (wherever natural resources will allow it) through centralized schemes due to economies of scale. Such a pattern of energy supply, whether renewable or not, will generally offer limited scope for meeting the development needs of rural areas, since energy distribution costs will continue to be high and users' incomes relatively low. As a result of these considerations, governments must consider the economic attractiveness of using small-scale renewable energy technologies for productive activities such as irrigation, crop drying, food processing or milling, as well as for household needs.

Many recent studies have analyzed the technical, economic and social aspects of a given technology,[11] but few have presented comparative economic evaluations of alternatives available for a given end-use. Even where these evaluations have been made, technology choice has been narrowly defined and methodologies adopted have varied from one author to another. As a result, a clear picture of the financial feasibility (from the users' viewpoint) and economic profitability (from the society's viewpoint) has not emerged. It is hoped that inclusion of these case studies will help to clarify the methodological aspects of comparative economic evaluation of renewable energy technologies.

Technologies and Countries Selected

In Chapter 4, human- and animal-powered lifts[12] are compared with diesel and petrol pumpsets for low-lift irrigation in Bangladesh and Egypt. In Chapter 5, windmills are compared with diesel and kerosene pumpsets in Sri Lanka, Kenya, Cape Verde, and the Sudan. Finally, in Chapter 6, a more comprehensive evaluation is made of a larger number of options for pumping in the Gangetic plains of North India, including: electric pumpset using electricity from the regional grid; diesel pumpset; dual-fuel pumpset using biogas or pyrolysis gas (from a gasifier); solar-photovoltaic and solar-thermal systems; and windmills.

There are important differences in the methodology used in these three chapters in terms of the range of technology choices considered; procedure of estimating required capacities; comparison of various alternatives; and the type of adjustments made to reflect economic costs of major inputs. The first two chapters (4 and 5) compare the technologies in terms of their costs per cubic meter (m^3) of water. Their findings are useful for policy making, further research, and, to some extent, farmers' decisions, but they do have some limitations. The main one is that they do not explicitly analyze the sensitivity of results to crucial assumptions regarding farm size, crop pattern, monthly availability of water, and capacity utilization of alternatives using conventional energy sources. These two studies are included to illustrate the current "state-of-the-art" for methodologies of comparative economic evaluation. These studies do not, however, offer the in-depth analysis required for understanding the range of available alternatives and their economic ranking. Taking these limitations into account, Chapter 6 therefore attempts to provide the type of analysis that should be carried out. In this case, the optional technologies are compared in terms of their net present values. Further methodological considerations emerging from the literature[13] and the following three chapters deserve some additional remarks.

Evaluation Methodologies

The microeconomic evaluation of alternative technologies for a given end-use can be done in financial terms, using market prices, or in economic terms, using shadow prices. However, since shadow prices for all different inputs cannot always be estimated, owing to lack of data and other reasons, intermediate methods are often used to arrive at approximations of economic costs and benefits. For example, major inputs such as fuels and foreign exchange may be shadow priced whereas other inputs may not. Such second-best methods have evident flaws, although overall shadow pricing based on guesswork has its drawbacks as well. Different methods are also used to estimate annual costs, particularly annualized capital costs. Beyond these general considerations, three approaches to financial and economic evaluation can be distinguished.

In the *first* approach, the financial and/or economic profitability of a technology is assessed by comparing the benefits and costs of using that technology. For example, the economic benefits from a solar-thermal system or a windmill are calculated in terms of the value of output of rice, maize, or wheat, by using the irrigation water supplied by the solar system or windmill. This method is termed as "viewed strictly in its own terms . . . without reference to competing systems."[14] However this method is not appropriate if the farmer (or the society) can choose between a solar system and a diesel pumpset to provide water, regardless of the additional benefits from irrigation. The introduction of a solar system would replace an existing (or potential) diesel pumpset. Thus the true benefit of the solar technology is the savings in resources (i.e., fuel, skilled labor, and possibly capital) that would otherwise have been used in the diesel pumpset. It is incorrect to attribute benefits from the use of irrigation water to the solar system, when in fact the same quantity and quality of irrigation water could be provided by competing technologies.

In the *second* approach, which is used in Chapters 4 and 5 (albeit with some differences), the cost per m^3 of water output of a technology is compared with that of an alternative technology (e.g., a windmill versus a diesel pumpset). If the unit cost of the technology in question is lower than that of the alternative being considered, it is concluded that the former is economic. One problem with this approach is that values for interest rates and life of equipment can always be chosen to favor a particular technology. Another problem is that it assumes that the technologies are comparable in terms of water output and utilization rate. A human-powered lift with a daily water output of 30 m^3 can be compared with a diesel pumpset with a daily output of 1,400 m^3 per day. However, it may not be possible to fully utilize available water because of a number of factors such as farm size; availability of water in the source (e.g., a dugwell), cropping pattern, seasonality of water output because of resource availability, which may not match demand (e.g., in the case of windmill); availability of fuel/electricity; equipment breakdown; and so on. For example, for a farm-size of 0.5 ha it would not be possible to use 1,400 m^3/day of water pumped continuously over a 63-day period. This may not even be possible or needed for a 14 ha farm, because of crop seasonality, rainfall, and water requirements. Hence a comparison between the cost per m^3 of the two devices, without taking the farm size explicitly into account, will give erroneous conclusions regarding the economics of a diesel pumpset. In this context, it is important to consider that some technologies can be used at a low level of output (e.g., human- or animal-powered lifts, solar-photovoltaic systems, and windmills). But some other technologies, such as diesel pumpsets, cannot be scaled down (even if they can be scaled down, the capital costs are reduced marginally). Under these circumstances, the possible under-utilization of diesel pumpsets should be explicity taken into account when comparing the economics of different technologies. Another alternative may be to include in the analysis technologies that can be scaled down (e.g., petrol or electric pumpsets) and then do the cost comparison. Even here it is necessary to specify whether the alternatives are considered under their "best" performance conditions or under average field conditions. There is a tendency to compare the techno-economic characteristics of one technology as it is expected to perform in the field against those of the alternative technology as it has *actually* performed in the field. A lack of clarity on these issues is likely to distort the comparison.

The financial analysis using market prices is sometimes modified by replacing market prices (usually subsidized) by economic costs (as reflected by world prices). In some cases, the shadow price of foreign exchange is used to reflect the scarcity premium. Most of the case studies summarized in Chapters 4 and 5 estimate annual savings in foreign exchange and employment created by technologies using renewable sources. In these cases, macroeconomic considerations are reflected by the estimated net savings in foreign exchange and employment creation by a specific program (such as the installation of 1,000 windmills).

In a *third* approach, which may be regarded as an improvement over the above and as used in Chapter 6, the economic evaluation and ranking of technologies is done in terms of cost-effectiveness criteria for "equivalent" irrigation systems. In this methodology, each individual technology is designed so as to provide the same reliability and performance over the period under consideration. These are "equivalent" in the sense that each one provides the same quantity and quality (time pattern) of water for irrigation, thus giving the same levels of agricultural benefit.

These systems are compared and ranked in terms of the sum of the present value of capital costs and operating costs incurred over a fixed period (e.g., 10 years). Since the life of each component of capital equipment may vary from one technology to another, replacement costs for some types of capital equipment must be included. Since the options are equivalent in terms of water output, the benefits from irrigation are the same. Thus, the alternative with the lowest present value of total costs is considered the best, that is, as giving the highest benefit/cost ratio. For a proper economic evaluation of various energy alternatives for irrigation pumping—as done in Chapter 6—it is necessary to:

1. Determine a few representative cropping patterns and estimate monthly water requirements for each crop rotation on the basis of agro-climatic conditions and effective rainfall
2. Analyze resource availability in the context of solar and wind energy, such as global solar radiation, wind speed and frequency (hourly, daily, and monthly). Apart from the magnitude and monthly distribution of these parameters, it may also be necessary to know the extent of "repeatability" of these magnitudes from one year to another
3. Identify the critical month for each crop rotation—the month that shows the highest ratio between monthly water requirement and monthly gross solar radiation (or monthly mean wind speed)
4. Estimate the installed capacities of equipment required to deliver the needed amount of irrigation water. This is done from the analysis of critical months and relevant energy resource data and requires the use of appropriate overall efficiency factors (e.g., solar to hydraulic) to match energy requirement for the critical month
5. Estimate the capital and operating costs of each system from the viewpoint of society. This may involve use of shadow prices for inputs such as electricity and diesel oil as well as for foreign exchange for the imported components of costs
6. Estimate the present value of the sum of these costs (for the purpose of comparison and ranking) by using an appropriate social discount rate for each system

It is only after such an in-depth analysis has been done that one can effectively address the policy issues that affect choice of technology by the farmers/consumers. Based on such analyses, one can outline the energy pricing and tax/subsidy policies that will promote private investments in channels that are found profitable from the society's viewpoint. The sensitivity analyses carried out with respect to technical parameters can also indicate areas to which R & D efforts should be directed. Many of these evaluations have been done for a particular technology, but very few studies are available where technologies have been compared and ranked for a given end-use. Moreover, adequate emphasis has not been given to the role of energy pricing policies (particularly subsidies for electricity and oil products) as a constraint to adoption of such technologies. Thus, there is need for a review of existing methodologies for comparative evaluation of various technologies.

LIMITATIONS OF THE VOLUME

It is neither possible nor desirable to cover all the important issues concerning the energy options addressed in this volume. Given the literature already available, such a task would be superficial and not very useful. Nevertheless, it is important to define the limitations of the volume.

First, there is some variation in the emphasis of the individual chapters due to their different authorship, data-related constraints, and so forth. This was unavoidable, even though considerable effort was made to modify some of the original papers so as to achieve an acceptable level of consistency throughout. To some extent, this limitation has been dealt with in the above sections, by presenting the chapters in a broader context of related issues.

Second, there are organizational and institutional aspects of planning and implementation that lie beyond the scope of socioeconomic analysis but are crucial in determining the success or failure of programs and projects involving the selected technologies. For example, organizational aspects of raw material supply, as well as interinstitutional linkages concerning ethanol production, oil-refining structure, transport policy, and energy pricing, deserve more discussion than given in Part I. These aspects have nonetheless been analyzed elsewhere in the literature on Brazil. For India, it would not be very useful to include a detailed analysis of these aspects if the macroeconomic implications of a fuel-ethanol program are still in question. Indigenous technological capacity, energy pricing and taxation, management training and technical cooperation, and other institutional aspects are also very important in the adoption of renewable energy technologies for irrigation. Except for energy pricing, these other aspects are not addressed in Part II, and certainly deserve much more attention in situations where technologies may be economically feasible. If their financial and economic feasibilities are promising or not yet settled, however, one may express doubts about the immediate usefulness of addressing training and other related aspects in detail.

Third, there are a number of other issues that are not discussed in the volume, either because they were not perceived as crucial to the questions addressed or because they lie outside the scope of ILO action. These include transfers of energy technology; financial assistance to energy programs; implications of market structure and trade policies for the choice of energy technologies; and mobilization of entrepreneurial capacity for renewable energy development.

In the light of the findings and limitations of the volume, the concluding chapter (Part 2) provides some brief discussion of implications for further research and training.

NOTES

1. See, for example, Chapter 2 of El-Hinnawi, Biswas, and Biswas (1983); UN (1984); IDRC (1986); Chapter 8 of Pereira, Ulph, and Tims (1987); Munasinghe and Schramm (1983); Auer (1981) and World Bank (1981a).

2. Vegetable oils and oils from shales and tar sands are technologically mature but generally too uncompetitive at recent oil-price levels. Synthetic fuels from coal and natural gas are still technologically constrained; compressed natural gas and producer gas are constrained by capital costs associated with their use; and methanol cannot yet be used in available combustion engines because of its health hazards. Given the range of energy alternatives for different end-uses and the recent prices of petroleum-derived fuels, ethanol is likely to be used mainly in substitution for chemicals and transport fuels. However, it may gain eventual importance as a fuel option for irrigation water pumping and other productive activities as well as a substitute for kerosene in households.

3. See CENAL (1983); Murthy (1980); and World Bank (1980b).

4. See Chapters 2 and 3, as well as World Bank (1980b) and Pereira (1986).

5. For details about ethanol projects in various countries, see in particular the World Bank/UNDP's Energy Sector Assessments carried out since 1981.

6. Though internationally less accessible, some important literature has appeared in Brazil. Major studies include CNPq (1978); PLANALSUCAR, University of Sao Paulo, and Mauá Technology Center (1981); and CENAL (1983). Pereira (1986) provides a detailed list and some analysis of these and other literature.

7. Essentially, the premium implies that traded goods are valued higher in relation to non-traded goods than if the official exchange rate were used. For methodological details see Squire and van der Tak (1975) and Little and Mirrlees (1974).

8. Explanations of this method are given in Chapter 2.

9. Throughout this book, the concepts of financial costs and benefits refer to the standpoint of private users and are measured at market prices.

10. The concepts of economic costs and benefits refer to the standpoint of society—the whole economy—and are measured at shadow prices. At places, because of data-related constraints, not all relevant inputs can be shadow priced and approximations of economic costs are used. Such approximations are nonetheless qualified where appropriate.

11. A short list of some recent publications that offer a general coverage of these aspects was presented in note 1, above. Various journals specializing on energy (see Bibliography) offer reviews of current literature. Among those, *Asset* (published periodically by the United Nations University and UNESCO) provides valuable abstracts of studies on renewable energy technologies. Some other important contributions are mentioned below and in individual chapters of this book.

12. To avoid confusion, we try to use consistent terminology throughout the book. The term "lift" is used for human- and animal-powered devices, since the technologies discussed do not involve pumps in a technical sense. The term "pumpset" refers to pump plus engine/motor (powered by diesel, petrol, kerosene, electricity, or a combination of diesel and biogas or diesel and pyrolysis gas). The term "system" refers to photovoltaic and solar-thermal devices, including solar arrays or solar collectors plus pump and other "balance of system" equipment. The term windmill includes the pump. Also the term "system" is sometimes used in a broader sense, but "pump", "system" and "engine" are never used interchangeably.

13. For a discussion of the methodologies and applications of social benefit/cost analysis to renewable energy projects, see, for example, deLucia and Bhatia (1980); World Bank (1981a); French (1979); Bhatia and Niamir (1979); Ghate (1980);

deLucia and Lesser (1980); Hurst (1984); and Lichtman (1984b).

14. See, for example, French (1979) for an application of this method to the economic evaluation of a 40 hp solar-thermal pump and a 5.5. kW solar-photovoltaic pump.

I

MACROECONOMIC EVALUATION OF ETHANOL PRODUCTION

2

Impacts of Ethanol Production in Brazil

Armand Pereira

Brazil's ethanol program (PNA) was launched in 1975 when the unit costs of ethanol fuels were substantially higher than those of petrol. This situation continued until the second oil shock in 1979/80, when the average barrel (bbl) of ethanol produced in Brazil became clearly economical. For this reason, ethanol production targets established in early 1979 were much more ambitious than those defined in 1975: production was about 3500 million liters (m.l) in 1980 and over 10,000 m.l in 1985. In addition, the production of hydrous-ethanol cars was initiated in 1979, reaching about 250,000 units in 1980 and 590,000 in 1983 (some vans, light lorries and small tractors fuelled by hydrous ethanol have been introduced; also the number of gasohol cars adapted to run on hydrous ethanol was more than 30,000 in 1980 and two or three times this number in 1983).

Ethanol, however, lost its economic competitiveness in 1983[1] and its production became a heavy financial burden in 1986. If investment decisions had been made purely on the basis of the relative costs of ethanol and petrol, PNA would not have been launched. But such decisions were based partly on the uncertainties in the market trends of both oil and sugar, and partly on other objectives of ethanol production including foreign currency savings, employment, income distribution, development of R & D

This chapter is based on A.F. Pereira: *Ethanol, employment and development: lessons from Brazil* (Geneva; ILO, 1986).

capacity and of the capital goods sector, as well as the objective of self-reliance. Besides these considerations, other reasons made ethanol production an attractive proposition for Brazil.

First, ethanol had already been produced and used as both fuel and non-fuel in earlier decades, and the technology was known. Second, being one of the world's major sugar-producing countries, Brazil had (a) sugar mills that could easily add ethanol distilleries to their facilities, using sugar molasses and/or sugar cane as feedstock and sugar-cane bagasse as fuel, at practically zero opportunity cost; (b) institutional arrangements to support production (mainly those of the sugar industry); and (c) a fairly developed capital-goods sector capable of providing all the technological equipment for ethanol production. Third, the low sugar prices in the international market between 1969 and 1974, and again between 1976 and 1979, encouraged interest in ethanol among sugar entrepreneurs, who saw this alternative as a means of generating the extra revenues they needed to revitalize their businesses. Fourth, Brazil's privileged situation in terms of land resources did not restrain the possibility of production of both sugar and ethanol. The latter could be produced from additional land, without necessarily interfering with the optimal supplies of sugar to meet the domestic and international demands for the product. Being a member of the ISA, Brazil had restrictions on sugar output, which fluctuated below her total sugar-mill capacity throughout the 1975–1980 period.

PNA has gone through four conceptually different stages. The first five-year period was marked by general optimism, fast growth of output, and relative smoothness in implementation. Between late 1980 and the end of 1981, PNA confronted an institutional crisis, brought about by a combination of factors, such as the decreased consumption of car fuel in response to the 1979/80 oil shock; surplus of petrol; low sales of hydrous-ethanol cars because of unfavorable economic conditions and some earlier technical problems in these cars; rumors within government circles and the media about shortages of sugar cane and ethanol; and skepticism on the part of the car industry in 1979/80 to commit its production lines to increased output of hydrous-ethanol vehicles. By 1982 these problems had been resolved and PNA entered a third period of adjustment and improvement. Evidently convinced of PNA's economic feasibility, the World Bank agreed in 1980 to finance a considerable part of the additional capital required to meet the 1985 target for ethanol production. Financial, fiscal, and price policies were re-defined in 1981. Planning linkages between the ethanol industry and other related sectors were strengthened. The use of ethanol as a chemical feedstock increased nearly fourfold from 1980 to 1985. Some modifications in the oil product slate of the oil-refining sector have been introduced since 1980 in order to increase the oil substitution effect of ethanol, and further adjustments were being planned in 1982–1985. Also in this period, more emphasis was given to problems of land use and landownership concentration (major producing states now have designated areas for new ethanol projects) as well as to research on (a) substitution of ethanol for diesel, (b) use of alternative crops such as wood, cassava, and sweet sorghum, and (c) ways of improving efficiency of distilleries (small-scale distilleries in particular). With the decline in world oil prices, however, PNA began to face severe financial pressures, which culminated with the price collapse at the beginning of 1986. This has brought another crisis period for PNA that is likely to continue so long as crude oil prices fluctuate below $25–30/bbl. In 1986 alone, the government spent about $700 million in subsidies to keep PNA afloat. But production cannot be stopped, because Brazil now has a large fleet of hydrous-ethanol cars that cannot run on petrol or gasohol. In 1986, ethanol production was just under 12,000 m.l and some 30 percent of this was a surplus for an unfavorable export market. Reducing production significantly has been a major policy issue, but not an easy one to decide upon, as the world oil market remains highly unstable with a long-term tendency toward the price levels of 1982–1985 (i.e., $25–32/bbl of crude oil). Cutting production would force private distilleries into bankruptcy, and this would jeopardize earlier agreements between the government and the producers, while leaving the government with a bundle of unpaid loans. Efforts are being made to decrease ethanol production and consumption, but this objective cannot be achieved in the short term. Meanwhile, a more realistic policy issue has been to assess the extent to which ethanol prices can be increased without significantly reducing consumption.

The experience just summarized above is unique in the world and it is clearly worth some in-depth analysis. As can be expected from any pioneering venture, PNA has had positive and negative aspects, which have become a focus of debate inside and outside Brazil. Much of this debate, however, has been based on superficial analyses of some aspects while ignoring others. With the recent decline in world oil prices, PNA has been widely criticized, but even before 1983, when ethanol was fully competitive with petrol, the debate on PNA's merits and flaws tended to be generally split between favorable and unfavorable views—both biased by value judgments resulting from the lack of adequate tools and data to assess the full range of PNA's costs and benefits.

The difficulties in measuring overall costs include uncertainties about (a) the total amount, origin, and real cost of public investments, both directly in ethanol-related infrastructure and indirectly in the form of subsidies to private investments; (b) indirect investments in oil-refining adjustments that may have been or will be required to increase the oil substitution effect of ethanol; and (c) the effects on such costs of a number of uncertain variables (e.g., wide variations in inflation and exchange rates observed since 1975, and variations in the gestation periods of investments; overlaps between ethanol- and sugar-related investments as regards agricultural production, R & D, and other government services; and of substitution of sugar cane for food and export crops, which may have some negative implications for food price inflation, balance of payments, and income distribution). On the benefits side, similar difficulties also emerge in the economic valuation of foreign exchange savings, of employment and income effects, and of spillover effects on other sectors. Due to the above constraints, it is practically impossible to quantify all costs and benefits in the context of both ex-post assessments for policy evaluation and ex-ante assessments for further planning purposes. Consequently, decisions on ethanol production in Brazil, as well as judgments as to its social profitability, will be based to some extent on qualitative criteria including political choices, which are naturally subject to controversy.

Whatever praise and criticism it may deserve, PNA is irreversible, owing to its repercussions on the pattern of demand for transport fuels and on the structure of both the vehicle-manufacturing and the oil-refining sectors in Brazil. Thus the best contribution the researcher can make to policy making audiences inside and outside Brazil is to assess and qualify PNA's impact with a view to providing information that may be useful for policy improvements. In the light of PNA's major objectives, the first section below provides a brief summary of PNA's impact at the end of its first five-year period (based on Pereira, 1986). PNA's effects on the balance of payments and employment are analyzed in the second and third sections, followed by policy-oriented conclusions in the fourth and final section.

SUMMARY OF IMPACT

By the end of PNA's first five-year period, the production of 22.6 m.bbl of anhydrous and hydrous ethanol in 1980 required less than 1 million hectares (m.ha) of harvested land, 207 distilleries of varying scale,[2] and a total investment of about $2000 million (mid-1980 prices), excluding land purchases. This is roughly equivalent to an investment of $0.57 per liter.[3]

This production had a net impact on the balance of payments of about $520 million, and created direct employment for 41,000 permanent workers (16,000 in industry and 25,000 in agriculture) and 83,000 seasonal workers (6500 in industry and 76,500 in agriculture). This implies investments of about $15,000 per worker or $22,000 per work-year, which are considerably lower than those required in most energy subsectors including other renewables and oil. Indirect employment is also significant but difficult to assess with any precision.

Against these achievements, however, PNA may have contributed to a concentration in personal and inter-regional income distribution over the first five-year period. These effects seem clear, although their extent cannot be rigorously assessed on the basis of available data. For example, ethanol investments, which were highly subsidized, had been largely concentrated in the richer regions; Sao Paulo, the richest state, had increased its share from 65.2 percent in 1975/76 to 72.9 percent in 1979/80. This trend was to some extent unavoidable in this first period, as the majority of the distilleries in operation were set up next to existing sugar mills, due to cost advantages. The locations of these so-called annexed distilleries were therefore predetermined by the sugar industry. Although sugar mills and ethanol distilleries were allowed by legislation ("Sugar-Cane Statute") to control up to 40 percent of their raw material requirements, most to them had increased this percentage through land purchases. The objective of such legislation was to avoid landownership concentration and permit a broader participation of small farmers in sugar-cane cultivation. However, the greater the distilleries' landownership, the lower their costs of production (raw material from external suppliers is purchased at prices controlled by the government). In 1981, some distilleries controlled over 80 percent of such requirements, while the national average was about 65 percent.[4] This increasing concentration in landownership normally involved the substitution of large plantations for smaller land plots. This process caused displacements of food staples,[5] leading to increased seasonality and migration of workers. It also caused the eviction of small farmers and subsistence peasants who held land plots under rentals, partnerships, and illicit forms of land tenure, although these evictions have not been quantified.

For 10,000 m.l in 1985 (2.8 times higher than in 1980 and requiring roughly a threefold increase in investments and harvest land), the net balance-of-payments impact may have been 2–4 times higher than in 1980, depending on price trends and on adjustments in the oil-refining sector. The direct employment effect may have increased by a factor of 2.5–3 (rather than 2.8) as a result of two offsetting variables: increased mechanization and the increasing share of the North/North-East (N/NE) macro-region, where labor coefficients have been considerably higher than in the Center/South (C/S).[6] The N/NE will have increased its share in total ethanol production from 15.8 percent in 1980 to about 30–35 percent in 1985. PNA's impact on inter-regional income distribution will have improved over that in the first five-year period, due to the increasing share of poorer regions in total production, hence in investments and government subsidies. There are two major reasons for this change. First, the possibility of annexing distilleries to sugar mills has been practically exhausted in the state of Sao Paulo and other richer states in the C/S. Second, most distilleries commissioned after 1981 are autonomous units (not linked to sugar production), which can be guided to locations designated for the ethanol industry, including poorer states in the NE. The impact on the distribution of personal income may have also improved from 1980 to 1985, as the increasing control over choice of project location may have reduced the extent of displacements of both crops and landless farmers.

Although this chapter cannot address all the above issues, the following sections include analyses of the impact of ethanol use on the balance of payments and direct employment.

IMPACT ON OIL IMPORT SUBSTITUTION AND BALANCE OF PAYMENTS

The supply of ethanol in Brazil has had a mix of the following consequences:

—import substitution of petrol and petrochemicals;
—exports of ethanol and petrol;
—import substitution of naphtha (made possible by a decrease in the oil-refining fraction of petrol and an increase in that of naphtha, part of which has been used for producing diesel).

Apart from the intended replacement of chemicals (e.g., 13 percent in 1980) and petrol, the other listed effects occur because of imbalances between the relative consumption and production of fuels, which are related to rigidities in the configuration of Brazil's oil-refining sector.

Patterns of production and consumption of oil products are very important for the planning of ethanol programs. These issues were analyzed in detail in Pereira (1986) but they cannot be given central attention in this brief discussion, particularly since the emphasis here is on the ethanol production impact rather than planning aspects. However, a few points need to be drawn from Table 2.1, which summarizes the data required for assessing the balance-of-payments impact of ethanol (Figure 2A.1 in the Appendix to this chapter shows surpluses and deficits of oil products based on detailed data on oil product balances for 1979–1982).

The table shows that, except in 1980 when a major change in the oil product slate occurred, there have been surpluses of petrol and ethanol (lines 32 and 34). The export market absorbs the petrol surplus fully but only a small part of the ethanol surplus; as a result, the unexported remainder must be consumed locally. This has been done by gradually increasing the percentage of anhydrous ethanol in gasohol, but since 1982 the government also had to promote consumption of hydrous ethanol.[7]
The following items must be taken into account in the assessment of PNA's balance-of-payments impact:

- c.i.f value of substitutions of ethanol for imports or fuels and petrochemicals (or their feedstocks, e.g., naphtha);
- c.i.f value of imported production inputs;
- cost of imported capital (and/or equipment and technology whenever the case)

for ethanol production as well as for investments in changes in the oil-refining structure due to increased ethanol supply;

- loss of export revenues from sugar (or manioc starch, or molasses, if these are used as raw material) in situations where (a) sugar exports can be increased, and (b) f.o.b export revenues per ton of sugar cane are higher than the c.i.f import cost of petrol (in amounts equivalent to the output of ethanol per ton of sugar cane) but where a captive domestic market for ethanol restricts higher exports of sugar;
- f.o.b value of ethanol exports;
- loss of ethanol export revenues in situations where world prices of ethanol become higher than those of petrol, and where a captive domestic market for ethanol, such as that of hydrous-ethanol cars, does not allow a reverted substitution of petrol for ethanol;
- f.o.b value of petrol exports, if any, made possible by the substitution of ethanol for petrol;
- c.i.f import cost of the exportable surplus of petrol (or of its crude oil equivalent), which is debited against the credited revenues from petrol exports (assuming these could only be possible as a result of surplus with direct or indirect implications for oil imports);
- foreign exchange losses derived from higher imports or lower exports of food crops, which are caused by the substitution of ethanol raw materials for these crops.

Any assessment of the balance-of-payments impact of ethanol production tends to be slightly distorted, because most of the items listed above cannot be measured accurately. These are briefly discussed below with reference to the impact in 1980, which is summarized in Table 2.2. Further considerations will then be made about the impact in 1982 (also in Table 2.2).

The c.i.f value of imported inputs to industrial and agricultural production were $170.4 million and the costs of imported capital were about $200 million.[8]

Losses in sugar export earnings were not obvious even in 1980 and the beginning of 1981, when sugar prices reached their highest levels since 1974 and the ISA quota was temporarily suspended. This deserves some explanation. Brazil was able to export as much sugar over the 1970s as the ISA allowed. Variations in sugar production (Appendix, this chapter) have been caused mainly by price factors and restrictions on supply rather than the expansion of sugar-cane production for ethanol. The author's review of Banco do Brasil's export accounts found sugar exports rising from $39.5 million in January 1981 to $181 and $114.3 million f.o.b in February and March—the latter was roughly 100 percent higher than sugar export earnings in March 1980, and the decrease from February to March 1981 was caused by the decrease is sugar prices. In 1980, Brazil had a sugar-mill

Table 2.1

Consumption (cons.) and Production (prod.) of Ethanol and Ethanol-related Oil Products in 1976–1983 and Effects of Ethanol on Oil Import Substitutions and Fuel Exports in 1980–1982 (m.bbl per year)*

	1976	1977	1978	1979	1980	1981	1982	1983
1. Total crude oil cons.[1]	336	345	386	407	398	385	381	-
2. Diesel cons.	86.5	92.8	101.3	110.3	118.5	116.3	117.9	-
3. " prod.	88.0	96.9	107.1	111.5	123.4	116.2	123.1	-
4. Naphtha cons.	-	-	-	20.9	22.4	25.3	30.6	-
5. " prod.	-	-	-	20.7	23.0	25.5	27.3	-
6. Additional naphtha cons. for diesel prod.[2]	-	-	-	5.7	6.8	7.5	7.8	-
7. " " prod. " " "	-	-	-	5.7	6.8	7.5	8.7	-
8. Ethanol cons. (total, 9+10+11)	3.3	6.9	11.8	18.0	19.9	18.4	25.6	34.7
9. Non-fuel ethanol cons.	2.0	2.5	2.3	2.1	3.0	2.5	2.4	2.5
10. Fuel anhydrous ethanol cons.	1.3	4.4	9.5	13.9	14.2	7.2	12.7	13.7
11. Fuel hydrous ethanol cons.	0	0	0	2.0	2.7	8.7	10.5	18.5
12. Fuel ethanol cons. (10+11)	1.3	4.4	9.5	15.9	16.9	15.9	23.2	32.2
13. Ethanol prod. (total, 14+15)	3.8	6.7	12.5	18.5	22.6	25.2	31.6	43.0
14. Anhydrous ethanol prod.	1.7	4.6	10.3	15.1	15.5	11.5	15.7	18.9
15. Hydrous ethanol prod.	2.1	2.1	2.2	3.4	7.1	13.7	15.9	24.1
16. Fuel ethanol prod. (13-9)	1.8	4.2	10.2	16.4	19.6	22.7	29.2	41.5
17. Petrol cons. (18+19)	91.9	85.0	86.8	84.8	73.3	69.8	66.8	-
18. Aviation petrol cons.[3]	0.5	0.6	0.6	0.8	1.0	1.1	1.0	-
19. Automotive petrol cons.	91.4	84.4	86.2	84.0	72.3	68.7	65.8	-
20. Petrol prod.	93.7	85.8	93.1	88.4	73.7	75.9	77.1	-
21. Gasohol cons. (19+10)	92.7	88.8	95.7	97.9	86.5	75.9	78.5	-
22. Total cons. of gasohol and fuel hydrous ethanol (21+11)	92.7	88.8	95.7	99.9	89.2	84.6	89.0	-
23. Total prod. of gasohol and fuel hydrous ethanol (20-18+16)	95.0	89.4	102.7	104.0	92.3	97.5	105.3	-
24. % of anhydrous ethanol in gasohol (10/21)	1.4%	5%	10%	14%	16.4%	9.5%	16%	-
25. Petrol prod. as % of gasohol cons. (20/21)	101%	96.6%	97.3%	90.3%	85.2%	100%	98%	-
26. Petrol prod. as % of total cons. of gasohol and fuel hydrous ethanol (20/22)	101%	96.6	97.3%	88.5%	82.6%	89.7%	86.6%	-
27. Total prod. of gasohol and fuel hydrous ethanol as % of its cons. (23/22)	102%	100%	107%	104%	103%	115%	118%	-

Effects of ethanol on oil-import substitutions							
Direct effects							
28. Petrochemicals (part of non-fuel ethanol)[4]	–	–	–	–	1	1	1.2
29. Petrol (22-(20-18))[5]	–	–	–	–	16.0	9.7	12.9
Indirect effects[6]							
30. Naphtha	–	–	–	–	2.3	0.3	0.6
31. Diesel (6)	–	–	–	–	6.8	7.5	7.8
Effects of ethanol on fuel exports							
32. Exportable surplus of petrol ((20-18)-(22-12))[7]	1.8	0.8	6.3	4.0	0.4	6.1	10.3
33. Exports of petrol				2.9	2.3	9.2	9.9
34 Exportable surplus of ethanol (8-13)	0.5	-0.2	0.7	0.5	2.7	6.8	6.0
35. Exports of ethanol	0.12	0.03	0.09	0.69	2.1	2.2	2.1

Note: Throughout this volume, the symbol "—" means "not applicable" or "not available."

[1]Based on details of oil product balances for 1979–1982 from MME (1983), which are summarized in Figure 2A.1, (Appendix). The consumption of the oil products listed excludes exports and includes (i) imports and (ii) storage, adjustments, and losses.

[2]Volume of crude oil consumed by domestic refineries to produce oil products.

[3]The additional production of naphtha for producing diesel was equal to the additional consumption except in 1982. The gap in 1982 was presumably because of storage and adjustments in 1981, when diesel production decreased, and probably because of the deficit of naphtha for petrochemical and fertilizer production.

[4]Aviation petrol is consumed by small private aircraft and it does not normally include fuel anhydrous ethanol.

[5]In 1980, the ethanol-based capacity for petrochemicals (i.e., ethylene, acetaldehyde, ethyl acetate, octanol, butanol, ethylene glycols, ethyl chloride, and ethyl ether) was 0.9–1.0 m.bbl according to experts at the Executive Committee for National Alcohol (CENAL) and Petroquisa (a subsidiary of PETROBRAS). Detailed time-series data have not been available in official publications, however.

[6]This is calculated by subtracting (i) petrol production minus aviation petrol consumption from (ii) the gasohol consumption plus the petrol-equivalent amount of fuel hydrous ethanol.

Notes to Table 2.1 (*Continued*)

[7] From 1979 to 1980, imports and total consumption of crude oil decreased and there were no naphtha imports. The increased production of naphtha (line 5) to meet the increase in consumption of the fertilizer and petrochemical sectors (line 4) was possible by virtue of the decrease in the fraction of petrol from 22.9 percent of each bbl in 1979 to 18.8 percent in 1980. This increased production can therefore be indirectly attributed to ethanol, since its supply made it possible to avoid separate imports of naphtha or fertilizers and petrochemicals. By the same token, the increased production of diesel through additional naphtha production (line 7) turns out to be another extended effect of ethanol, although this is strictly a consequence of the peculiar features of Brazil's fuel supply/demand balance and of ongoing adjustments in the oil-refining sector. In 1981/82, these effects were less evident, judging from the data in this table and more disaggregated MME data on oil product balances. The indirect substitution effect of ethanol for naphtha in 1982 is estimated as the difference between (i) the increased production of naphtha from 1981 to 1982 and (ii) the increased production of petrol from 1981 to 1982 (i.e., 1.8 − 1.2 m.bbl = 0.6 m.bbl); the same procedure was used for 1981. The logic of this method is this: had there been no shift from petrol to naphtha, the increased output of petrol and naphtha would have been the same. These indirect effects, however, should not be taken into account in the balance-of-payments analysis since, for a given level of ethanol supply, decreases or increases in petrol production (which affect the output of naphtha) are already directly reflected in decreased imports or increased exports of petrol, under the method used.

[8] The petrol surplus represents the amount of petrol available for automotive purposes minus the total fuel ethanol from the total consumption of gasohol and fuel hydrous ethanol; the result is the same if, instead, one subtracts fuel anhydrous ethanol from gasohol consumption. The logic of this calculation is this: as fuel ethanol production increases faster than the consumption of fuel ethanol and automotive petrol, there is a reduced need for petrol to meet this consumption.

Source: Elaborated from official data by the National Petroleum Council, Ministry of Mines and Energy, Institute of Sugar and Alcohol, and the Export Bureau of Banco do Brasil.

capacity of about 10–11 million tons (m.t) and produced less than 8 m.t of sugar, having exported 2.5 m.t at $504/t f.o.b ($13,000 million). In principle, the ISA would have allowed a Brazilian export of 2.2 m.t, but the quota was suspended in February 1980, as world sugar prices surpassed $0.23/lb ($480/t). This may lead one to think that Brazil may have been able to export more sugar in 1980, as was, in fact, suggested by Islam (1980). But, even if this were true (and there is no confirmation), the short-term marginal benefits from these additional exports would have pushed world sugar prices down and possibly reduced export revenues in following years. The additional 130 million tons of sugar cane required to meet the ethanol production target in 1985 would represent an additional production of 12–13 m.t of sugar, which would be nearly 50 percent of the world sugar trade. This would mean an increase in Brazil's share in this trade from

Table 2.2
Impact on the Balance of Payments: 1980 and 1982 ($ millions)

	1980				1982			
	Case A		Case B		Case A		Case B	
Item	Credits	Debits	Credits	Debits	Credits	Debits	Credits	Debits
1. Import substitutions [1]								
-Petrochemicals	41	-	41	-	44.9	-	44.9	-
-Petrol	707.2	-	707.2	-	522.4	-	522.4	-
2. Imported inputs								
-Diesel		77.8	-	77.8	-	108.8	-	108.8
-Fertilisers	-	57	-	57	-	79.7	-	79.7
-Insecticides, herbicides and fungicides	-	27.8	-	27.8	-	38.9	-	38.9
-Industrial inputs	-	7.8	-	7.8	-	10.9	-	10.9
3. Losses of sugar exports	-	0	-	0	-	0	-	0
4. Cost of imported capital for PNA's investments	-	200	-	200	-	279.6	-	279.6
5. Ethanol exports[2]	117.6	-	151.2	-	117.6	-	117.6	-
6. Losses of ethanol exports	-	0	-	0	-	0	-	0
7. Petrol exports	16.8		16.8	-	394.5	-	394.5	-
8. Import cost of petrol exports[3]	-	14.3	-	0	-	293.5	-	0

(Continued)

9. Cost of imported capital for oil-refining investments due to ethanol	-	Fcl	-	Fcl	-	Fc.1	-	Fcl
10. Foreign-exchange losses from from crop substitutions	-	Fc2	-	Fc2	-	Fc.2	-	Fc2
TOTAL NET BALANCE		+497.9-(Fcl+Fc2)		+545.8-(Fcl+Fc2)		+268.0-(Fcl+Fc2)		+561.5-(Fcl+Fc2)

Note: Oil product import savings and imported inputs to production are valued in c.i.f prices, and exports in f.o.b prices. Oil product f.o.b prices (per bbl) used are annual averages (Rotterdam spot market as quoted monthly in the *Petroleum Economist*); for naphtha, diesel, and petrol, these were respectively $38.8, $41 and $42 in 1980, and $35.4, $38.6, and $38.3 in 1982. Petrochemicals are valued at c.i.f prices of naphtha, which is their major feedstock. The c.i.f prices of these oil products are increased by a shipping cost of $2.2 in 1980 and an estimated $2.0 in 1982. Input prices are from UNCTAD's *Yearbook of Trade Statistics.* Data on import substitutions (line 1) and exports (lines 5 and 7) are in Table 2.1. Data on imported inputs (line 2) are in note 8 at the end of the chapter. Other items are discussed in the text. The differences in assumptions in Cases A and B for 1980 and 1982 are stated briefly in the table notes below and are also discussed in the text. Fc1 and Fc2 are symbols for the costs in lines 9 and 10, which cannot be estimated on the basis of available data (see text). Ethanol production was 22.6 m.bbl in 1980 and 31.6 m.bbl in 1982.

[1] Indirect effects on import substitutions of naphtha and diesel (Table 2.1, lines 30 and 31) are not credited, since these are already reflected in the petrol import substitutions.

[2] For 1980, Case A includes exports taken place in the year only (2.1 m.bbl), and Case B includes the exportable surplus (2.7 m.bbl). For 1982, both Cases A and B include observed exports only (also 2.1 m.bbl), since the exportable surplus in 1981 and 1982 was about three times higher than exports. Ethanol export prices were about the same in 1980 and 1982: $56/bbl.

[3] For both 1980 and 1982, Case A assumes this cost at the margin; it includes the c.i.f cost of crude oil ($32.2 f.o.b plus $2.2 shipping cost per bbl in 1980, and $25.5 f.o.b plus $2.0 in 1982), increased by foreign exchange component of the petrol-refining cost ($1.3 in 1980 and $1 in 1982—at borrowing conditions assumed in the second section of this chapter (note 8) for domestic refining costs of $10 in 1980 and an estimated $8 in 1982; see text and note 10), totalling $35.7 in 1980 and $28.5 in 1982. Case B values the import cost of petrol exports at zero, assuming the demand for different oil products made it economical to import the crude oil that produced the exportable surplus of petrol.

Source: See Notes at end of this chapter.

less than 10 percent in 1980 to about 40–50 percent by 1985. Even assuming this increased share would be allowed, world sugar prices would be substantially reduced. Thus, it may be argued that Brazil's commitment to ethanol may sometimes contribute to higher sugar prices.

There is also no reason to believe that there were losses in exports of molasses, which accounted for 25 percent of ethanol production in 1980. Exports increased from $10 to $24 and $29 million f.o.b in the first three months of 1980, and from $16 to $30 to $41 million during the same months in 1981. Provided that the ISA is able to maintain control over sugar supply, and that effective short-term forecasting of demand for molasses and sugar is continued, Brazil is not likely to incur any considerable loss in export revenues from these commodities in the near future. This is because of the relatively large supply of petrol, as well as the possibility of (a) increasing sugar production, (b) changing the relative share of molasses and sugar cane in ethanol production, and (c) reducing supply of anhydrous ethanol for gasohol blends. Besides, there is a slight risk that the ISA may become less effective in the future if nonmember OECD producers continue to increase their share in sugar supply. If this happens, world prices will probably drop considerably, and the opportunity cost of sugar cane for ethanol will decrease as well.

Losses of ethanol export revenues are uncertain because of fluctuations in world demand and prices. Moreover, it is unlikely that any such losses were incurred at least until 1983, judging from the recent trends of supply and consumption of petrol and ethanol, and again the possibility of reducing the percentage of anhydrous ethanol in gasohol blends. In 1980, ethanol exports increased by 219 percent over 1979, with the average bbl priced at $56 f.o.b ($16 above the average price in 1979) leading to a revenue of $117.6 million (2.1 m.bbl) compared to $26 million and $3.5 million in 1978 and 1979, respectively. Export losses may nevertheless occur in future years, due to (a) the captive market of hydrous ethanol for cars running exclusively on this fuel rather than on gasohol; (b) the differentiated international demand for hydrous and anhydrous ethanol; and (c) the constant fluctuations in prices and demand for these products. Although the average export price was $56/bbl in 1980, prices reached short peaks of $80/bbl. During the first two months of 1981, prices rose to $79.5/bbl, but by August 1981 the volume of exports were the lowest since early 1979 and prices decreased to less than $55/bbl, having fluctuated within the $50/bbl range at least until mid-1983. Table 2.1 shows that annual exports of ethanol have been much lower than the exportable surplus. This raises a methodological controversy as to whether Brazil would have been able to export the remaining surplus from 1980 in the following year.

Petrol surpluses have been rising substantially since 1980, but have been easily exported. For balance-of-payments analysis, one must take the exportable surplus into account rather than the exports that actually took place in a given year, since the latter are not all necessarily related to ethanol. Even in recent years, there have been some very minor imports of petrol, but these may perhaps consist of high-octane petrol or may be a result of trade agreements. However, the real resource cost of petrol exports cannot be estimated without a detailed study of the oil-refining sector.[9]

For notional purposes, Table 2.2 presents two cases for 1980 and 1982, under different assumptions as regards the import cost of petrol exports (line 8) and ethanol export revenues (line 5). *Case A* is less favorable than *Case B.* It values the import cost of petrol exports at the margin, under the assumption that it was economic for Brazil to import separate supplies of LPG and other oil products (i.e., diesel and possibly kerosene) that would have been in deficit if crude oil consumption had been decreased below the 1980 and 1982 levels (hence reducing surpluses of petrol and fuel oil). This import cost is calculated (per bbl) as the c.i.f cost of crude oil (average Rotterdam spot f.o.b price of $32.2 plus shipping cost of $2.2 in 1980, and $25.5 plus $2 in 1982), increased by the foreign exchange component of the refining cost of petrol,[10] totalling $35.7 in 1980 and $28.5 in 1982. In Case B, it is assumed that it was more economical to import crude oil than to import separate supplies of the above products, and no import cost is debited.

As regards ethanol exports, different sets of assumptions are made for 1980 and 1982. For 1980, Case A credits exports taken place in that year only (2.1 m.bbl, which was less than the exportable surplus), under the assumption that Brazil continued to find bottlenecks for ethanol exports and had to increase domestic ethanol consumption through further incentives to Sunday drivers. This assumption can be justified on the basis of the increasing gap between the exportable surplus and the observed exports (Table 2.1, lines 21 and 22). Case B assumes that the remaining surplus (0.6 m.bbl) was exported in 1981. For 1982, the observed ethanol exports (2.1 m.bbl) are credited in both Cases A and B, since it is evident that the remaining exportable surplus (3.9 m.bbl) would hardly have been exported in the following year (the remaining surplus in 1981 was even higher).

For 1980, the difference in PNA's net balance of payments based on either set of assumptions is not significant: $497.9 million for Case A, and $545.8 million for Case B. In either case, the impact is significantly positive, especially when considering the limited amount of investments that PNA required for the infrastructure of its first five-year period.

This estimated impact omits foreign exchange losses (symbolized as Fc2 in Table 2.2, line 10) that may have been caused by substitution of sugar cane for food and export crops. Although these cannot be estimated on the basis of information available, their extent, if any, will have been very minor and mainly associated with short-term lags in crop relocations.[11] It also omits costs of imported capital (Fc1 in line 9) that may have been required for ethanol-related changes in the oil-refining sector in 1980/81. An estimation of these costs requires a very detailed assessment of the refineries, which is beyond the scope of this study.[12] Nevertheless, these costs should not significantly change the above findings, since changes in the oil-refining structure relate not only to ethanol and petrol, but also to fuel oil, diesel, and other products. Furthermore, ethanol production may also avoid investments in the oil-refining sector, under certain fuel supply/demand conditions.

PNA's balance-of-payments impact in 1980 was unusually favorable, however. As changes were introduced in the oil refineries, petrol production was at its lowest level since the early 1970s. This allowed a relatively high substitution of ethanol for oil imports as compared with 1981 and 1982.

On the basis of the data in Table 2.1 and the assumptions made for 1980 regarding sugar exports and crop substitutions, PNA's impact on the balance of payments in 1982 may be estimated at $268.0 million for case A and $561.5 million for case B. For this estimate, 1982 average Rotterdam spot prices of oil products were also used; import costs of inputs, except diesel, were increased proportionately to the increase in ethanol output, assuming no changes from 1980 to 1982 in input coefficients and their c.i.f prices; for diesel, no change in coefficients was assumed, but 1982 c.i.f prices were used since these were considerably lower than in 1980; and financing costs of imported capital were also increased proportionately. This method and assumptions for estimating costs of imported industrial inputs, fertilizers, and capital may imply some distortion, but do not change the final figures significantly.

In years in which there is a large surplus of petrol (as in 1982), the major item that makes a big difference to PNA's balance of payments is the import cost of the petrol surplus. Although neither this cost nor the overall economic cost of petrol had been determined by the government until late 1983 (see note 9), it is imperative to do so in the light of recent trends in consumption and production of fuels. Unless such costs are known, it is impossible to reach firm conclusions about the net economic benefits of exporting petrol and, therefore, of having an increasingly large output of ethanol. By the same token, the comparative benefits of the oil sector options discussed earlier cannot be clearly assessed.

Given the increasing surpluses of car fuels, and also of fuel oil [which, according to CENAL (1983, p. 21) has been exported at 61–68 percent of the f.o.b price of crude oil, depending on sulphur content], the assumption made about the petrol import cost in Case A is much more likely to be correct than that in Case B. If so, then it may be argued that ethanol output has grown beyond the optimal level. Table 2.1 shows car fuel consumption lagging well behind production (see line 27) in spite of the price and fiscal incentives given to ethanol cars since 1982. Further incentives in 1982 and

1983 began to reduce the ethanol surplus after 1982, but the overall benefits of this sort of policy are questionable, insofar as ethanol continues to be consumed primarily by cars for private use. Reducing the growth rate of ethanol output has not figured among the options preferred by the government, but it may deserve attention as well.

In principle, PNA's balance-of-payments impact may have improved by the end of the second five-year period as a result of increased direct substitution of ethanol for petrochemicals and shifts in consumption from diesel to gasohol and fuel hydrous ethanol. In practice, this depends on the production and surplus of petrol, on uncertain import costs of this surplus, and on the extent to which the oil-refining sector, together with demand management policies, will have succeeded in solving the petrol/diesel imbalance referred to earlier.

DIRECT EMPLOYMENT IMPACT

Ethanol production generates direct employment primarily in the operation of distilleries and in the cultivation and transport of sugar cane. Indirect employment is also created in the manufacturing of equipment for ethanol and sugar-cane production; the installation and construction of distilleries and of adjacent infrastructure; the operation of gasohol-mixing centers; the distribution of hydrous ethanol; R & D; government services; and other activities. Direct employment can be estimated on the basis of surveys of distilleries and plantations, but indirect employment can hardly be assessed because it includes mostly short-term jobs, and many of these are not exclusively tied to ethanol. For example, major manufacturers of distillery equipment also produce other types of equipment; jobs created in the manufacturing of hydrous cars or in the alcohol–chemical industry may be credited to PNA, but these jobs would have existed if petrol cars or petrochemicals were produced instead. Hence, this brief discussion focuses on direct employment only. A few introductory remarks will help clarify the employment data below.

In Brazil, ethanol is produced directly from sugar cane, but also indirectly from molasses in the case of annexed distilleries. Most distilleries operate for 150–180 days per year, 24 hours per day, in two 12-hour shifts during the harvest season. In the off season (150 days), agro-industrial activities encompass equipment and vehicle maintenance, construction, and so forth, whereas agricultural activities comprise crop renewal and maintenance.[13]

Based on the author's survey[14] in early 1981, Table 2.3 shows the year-round employment and the seasonal employment in individual autonomous distilleries of 90,000, 120,000, and 240,000 liters per day (lpd), according to the following occupational groups:

i. directors and top management;

ii. skilled personnel (i.e., agronomists, accountants, chemists, and other technicians);

iii. semi-skilled personnel employed in office services (i.e., secretaries, maintenance employees, etc.);

iv. semi-skilled people in agro-industrial operations (i.e., drivers, vehicle and equipment mechanics, work supervisors, etc.);

Table 2.3
Administrative and Agro-Industrial Employment in Autonomous Distilleries of 90,000, 120,000, and 240,000 lpd

	Scale of distilleries (lpd)								
	90,000			120,000			240,000		
Nature of employment/ Occupational groups	No. of jobs	(%)	jobs/ m.l	No. of jobs	(%)	jobs/ m.l	No. of jobs	(%)	jobs/ m.l
Administrative/ Agro-industrial[1]									
Year-round I and II	7			7			7		
Year-round III	10			10			11		
Year-round IV	27			30			40		
Seasonal IV (harvest season)	6			7			10		
Year-round V	27			30			50		
Seasonal V (harvest season)	22			24			40		
SUB-TOTAL year-round	71	72	5.2	77	71	4.3	108	68	3.0
SUB-TOTAL seasonal	28	28	2.1	31	29	1.7	50	32	1.4
TOTAL (7 and 8)	99	100	7.3	108	100	6.0	158	100	4.4

Note: These are approximate figures, based on the author's survey (see note 14, this chapter). Because of variations in scale and in capacity utilization (CU) of distilleries of the same scale, only 13 of the 28 autonomous distilleries surveyed could be taken into account. These included 8 units of 120,000 lpd, 2 of 90,000 lpd, and 3 of 240,000 lpd. Even among these, the 120,000 lpd units had CU variations of 14 percent, two of them with an output above 120,000 lpd. The standard deviations of the figures in last row, "No. of jobs" columns, were as follows: 8.00 (90,000 lpd), 8.89 (120,000 lpd) and 13.88 (240,000 lpd).
[1]Occupational groups are described in the text.

Source: Author's survey.

v. unskilled industrial workers;

vi. unskilled agricultural workers and farm-machine operators.

Seasonal employment is observed mainly in the group of industrial unskilled labor (v); much less so among agro-industrial, semi-skilled workers (iv); and is nonexistent in occupational groups i–iii, although some professionals in group ii (i.e., chemists, accountants, physicians) do not always have full-time jobs. In these administrative/agro-industrial categories, total employment (including year-round and seasonal jobs) amounts to 99 for an annual production level of 13.5 m.l; 108 for 18 m.l, and 158 for 36 m.l, thus showing the effect of scale on the labor coefficients per m.l.

Although annexed distilleries are predominant in Brazil (accounting for 82.5 percent of total production in 1980/81), they cannot be used as a basis for estimating employment effects of ethanol, because of overlaps in sugar- and ethanol-related operations, and because of their partial production of ethanol from molasses, which does not require agricultural activity. The demand for labor in annexed distilleries of 120,000 lpd, compared with that in autonomous units of the same capacity, is about 20–30 percent less in the administrative/agro-industrial categories, and lower by varying amounts in agriculture, depending on the share of production derived from molasses.

Table 2.4 shows the impact of individual autonomous distilleries of 90, 120, and 240 thousand lpd on *agricultural employment* in the C/S macro-region. The data are based on labor coefficients periodically researched by COPERSUCAR since 196/77: 0.145 man-days in the off season, and 0.232 man-days in the harvest season. The coefficients refer to an average of situations involving three to four harvest cuts from each planting. Off-season operations are semi-mechanized, and harvest operations include mostly manual cutting and mechanized loading.[15]

The demand for agricultural labour per m.l of ethanol is estimated at 13.7 jobs in the off season and 22 jobs in the harvest season, independently of the distilleries' capacity.[16]

Direct Employment in 1980 and Projections for 1985

Total direct employment cannot be estimated with precision because of variations in demand for labor among types and scales of distilleries and regions. Nonetheless, this employment is estimated here on the basis of previous data on autonomous distilleries, and of agricultural labor coefficients in the C/S macro-region.[17]

Table 2.4
Direct Agricultural Employment (unskilled workers and farm-machine operators) Created by Autonomous Distilleries: C/S Macro-Region

	90,000 lpd		120,000 lpd		240,000 lpd	
	Jobs	Jobs/ m.l	Jobs	Jobs/ m.l	Jobs	Jobs/ m.l
Harvest season (150-180 days)	298	22	397[1]	22	794	22
Off-season (150 days)	186	13.7	248[2]	13.7	496	13.7

Note. Data based on labor coefficients by COPERSUCAR, assuming a harvest operation of 150 days.

[1]120,000 lpd × 150 days = 1.8 m.l ÷ 70 liters per ton (1/t) = 257,000 tons × 0.232 man-days = 59,642 man-days ÷ 150 days = 397.

[2]120,000 lpd × 150 days = 18 m.l ÷ 70 lpt = 257,000 tons × 0.145 man-days = 37,265 man-days ÷ 150 days = 248.

Source: Elaborated from COPERSUCAR data.

For the production of 3500 m.l in 1980, total direct employment in the administrative/agro-industrial categories may be calculated by multiplying the number of jobs in each 120,000 lpd distillery by 207 commercial units in operation. This leads to figures (not rounded, to facilitate cross-references) of 15,939 year-round and 6417 seasonal jobs. For a production of 10,700 m.l in 1985/86, the figures increase to 48,728 year-round jobs and 19,618 seasonal jobs, assuming no changes in technology (Table 2.5).

Using the aforementioned labor coefficients for the 50 m.t of sugar cane required in 1980 (assuming that all ethanol was produced from sugar cane), agricultural employment is estimated to have been 48,333 jobs (for about 150 days) in the off season, and 77,333 jobs (150–180 days) in the harvest season (Table 2.5). For 1985/86, the corresponding figures are 147,700 and 236,419, assuming the same yields, techniques, and interregional distribution of production. The employment data for 1985 may be higher or lower than this estimate, depending on the net quantitative effect of two offsetting factors: increased mechanization and the increasing share of the N/NE regions.

What was the total workforce employed under PNA in late 1980 and early 1981? To estimate this, we need to know the breakdown of year-round (permanent) and seasonal jobs. We have such a breakdown for administrative/agro-industrial activities but not for agriculture. However, CENAL estimated that 68 percent of the jobs in agriculture were seasonal

Table 2.5
Direct Employment Impact of PNA in 1980 and 1985 (according to number of jobs in all occupational groups during the harvest season and the off season)

	1980	1985
Administrative/Agro-industrial[1]		
Harvest season	22,356	68,345
(Permanent jobs)	(15,939)	(48,728)
(Seasonal jobs)	(6,417)	(19,618)
Off-season	15,939	48,728
(Permanent jobs)	(15,939)	(48,728)
(Seasonal jobs)	(-0-)[3]	(-0-)
Agricultural[2]		
Harvest season	77,333	236,419
Off-season	48,333	147,761

Note: Figures for 1980 are based on the calendar year's production of 3,500 m.l and yields of 50 t/ha/yr and 70 1/t. Figures for 1985 are based on a 10,700 m.l production, assuming the same yields, labor coefficients, and techniques registered in 1980/81. Occupational groups are described in the text.

[1]Based on the author's field survey (see note 14, this chapter).

[2]Based on coefficients by COPERSUCAR for Brazil's C/S macro-region.

[3]Normally the case, although a few distilleries surveyed employed some temporary labor in the off season.

Source: See above notes.

and 32 percent were permanent. On average, these percentages seemed compatible with information gathered by the author in early 1981. Applying these figures to the number of jobs shown in Table 2.5, it is possible to estimate the total workforce employed under PNA by the end of its first five-year period, with a detailed breakdown of year-round (permanent) and seasonal (150 days) workers during both periods of the year (Table 2.6).[18] This amounts to about 40,000 and 82,000 permanent and seasonal workers, representing respectively 33 and 67 percent of the total.

These benefits should be considered against the employment implications of the substitution of sugar cane for other crops and pastures. The net impact depends on the quantitative significance of these displacements, and on the relative labor coefficients of the crops and pastures. Although reliable documentation was available for areas accounting for only 65 percent of production, it is nevertheless possible to draw some conclusions. On the basis of the labor coefficients for the crops and pastures displaced (excluding castor oil, for which no coefficient was found), it can be es-

Table 2.6
Total Workforce Directly Employed Under PNA in 1980

	Harvest season	Off-season	Year
Administrative/Agro-industrial			
Year-round workers	15,939	15,939	15,939
Seasonal workers	6,417	-0-	6,417
Agricultural			
Year-round workers	24,746	24,746	24,746
Seasonal workers	52,587	23,587	76,174
Total year-round workers	40,685	40,685	40,685 (33%)
Total seasonal workers	59,004	23,587	82,591 (67%)
Total year-round and seasonal workers			123,276 (100%)

Note: This assumes a composition of 68 percent seasonal and 32 percent permanent agricultural workers during the harvest.

Sources: See Table 2.5.

timated that over 15,650 worker-year jobs, or alternative compositions of year-round and seasonal jobs, were lost. This displacement implied a net gain of 24,450 worker-years of employment because sugar cane is roughly seven times more labor-intensive than pastures, and among the displaced crops (rice, maize, cotton, manioc, beans, peanuts, oranges, and castor oil), only cotton, peanuts, and manioc have higher labor coefficients per ha. Moreover, as we had argued in the first section of this chapter, the displaced crops and pastures may have been relocated to other areas and created employment elsewhere (see notes 5 and 11 at the end of this chapter).

Qualitative Aspects of Employment

PNA's employment impact is undeniably high; however the figures reveal a high seasonality of this labor. Furthermore, job creation in rural areas is not always a good indicator of improvements in the incomes of poor groups, much less of decreases in the disparities of personal incomes.[19] For example, the above displacements of crops and pastures create more jobs, but also more seasonality and migration of labor, and more landlessness. Although numbers cannot be estimated, field interviews with managers of plantations—as well as some recent literature[20]—suggest that many jobs created by PNA implied evictions of small farmers and subsistence peasants from land that they illicitly occupied or rented, but did not own. While these issues cannot be adequately analyzed in this paper,[21] some limited discussion is in order.

In Brazil, and in some other countries, the seriousness of the negative aspects of seasonality and migration of labor depends on the types of seasonal workers concerned and their degree of underemployment.[22] Some of PNA's seasonal workers were once itinerant migrants who settled in certain communities, while working for one or more proprietors in the vicinity. These workers invariably have their "employment books" signed, receive legally stipulated wages and benefits, and are protected by labor laws regulating employer/worker relations. Another group of seasonal laborers comprises youngsters, women, and older men—who are not part of the regular workforce. These are not typical migrants, either, since they usually live in nearby rural communities. Although some of them receive wages below the minimum rate, few of them seek permanent jobs or depend on cash incomes for subsistence throughout the year. In the N/NE states in particular, where ethanol production takes place mainly along the coast, a large part of the seasonal labor force during the harvest period consists of male workers who migrate temporarily from their small plots in the hinterland, to which they return with some savings after the harvest. A third category of seasonal workers comprises those who alternate between urban and rural jobs, although they are relatively few. And last, there is another group normally classified in Brazil as itinerant workers, with no fixed residence, who are hired mostly by labor contractors on a short-term basis. Most of these workers are not protected by labor legislation concerning employer/worker relations, including fringe benefits, although most of them earn at least the legal minimum wage. The emergence of labor contractors, operating as intermediaries for the plantations and normally without a social or economic interest in making optimal use of labor, has been a major factor preventing improvements in the standards of living for this last group of workers, who comprise a very large part of the workforce in the state of Sao Paulo.

The income-related disadvantages of itinerant workers are associated with their (a) job insecurity, (b) exclusive dependence on low wages, and (c) inability to generate nonmonetary income, often referred to as income "in kind," such as that provided indirectly by subsistence production of food and social assistance.[23] An increase in permanent jobs would tend to have a positive effect on these workers and, to a lesser extent, on those who alternate between urban and rural jobs, as opposed to other types of seasonal workers who rely on other sources of income.

It is impossible to document with precision the impact of PNA on personal incomes, because of the lack of data on the previous incomes of workers and the losses of income "in kind" (which may have been caused by landownership concentration, crop displacements, and their consequences). By the same token, the social effect of changes in personal incomes is even more difficult to assess, due to the lack of data concerning the previous marginal product of labor and the net social cost of increased consumption or of reduced leisure.

Wages have different implications for the incomes of different groups of workers employed under PNA. The data collected by the author in 1981 (see note 14) on wages of semi-skilled and unskilled workers (accounting for about 95 percent of PNA's workforce) suggest that the most significant improvements in incomes have taken place among semi-skilled agro-industrial and agricultural workers, such as lorry drivers, workers in vehicle and equipment maintenance, farm-machine operators, and other workers supervising industrial equipment. Persons in these occupations usually hold permanent jobs, although their work can scarcely be regarded as fulltime in the off season. This allows many of them to have some additional income-generating activities during this period. Their skills and limited supply have had a favorable effect on fringe benefits and wages. The wages of truck drivers and farm-machine operators, for example, ranged from $145 to $250 per month (in May 1981, at the shadow exchange rate) and varied substantially within each region, but were generally higher than those of similar jobs in large urban centers. In many distilleries, workers in these two occupations received overtime earnings during the harvest season, when their working hours increased to 10–12 hours per day, but the earnings also varied from one distillery to another. However, the income increase for this group of semi-skilled workers was not very representative of the general state of affairs, because their numbers were relatively few.

Unskilled workers with permanent jobs in industry and agriculture may have benefited slightly, not so much in terms of wage earnings but more in terms of the security of their incomes and job stability, taking into account the vicissitudes of Brazil's economy. On average, their wages were about the same as those in comparable activities in other sectors (minimum salary base of $50–63 per month in May 1981, at the shadow exchange rate). Overtime earnings during the harvest season may slightly improve incomes, but no consistent pattern was found in the author's survey.

For seasonal agricultural workers, still the largest group of employees, the findings are less encouraging. Most of these people worked 10–12 hours per day during the harvest, but earned average wages of about $63–70 per month (slightly above the minimum wage level, in May 1981, at the shadow exchange rate) and, in some cases, women and teenagers earned even less. These wages were a convenient source of extra income for rural families rooted in communities adjacent to sugar-cane plantations or for workers who migrated temporarily away from their family plots. But they were far too low (given Brazil's cost of living) for itinerant workers who depended exclusively on these wages to purchase food and other basic commodities. The fact that the latter workers earned little more than the legal minimum wage[24] hardly suggested an improvement in their incomes, because many of them (objective figures are not known) had secured wage-employment in replacement of some or all of their income "in kind."

CONCLUSIONS AND POLICY IMPLICATIONS

PNA's impacts on the balance of payments and employment at the end of its first five-year period are undeniably significant, given the relatively low investments that it required (see summary of impacts in the first section of this chapter). This chapter nevertheless suggests a number of problems that need policy attention.

1. The surplus of petrol since 1978 and the widening gap between production and consumption of both fuel ethanol and petrol since 1980 suggest that production grew too fast. While the increasing surpluses of petrol can be exported and those of fuel ethanol can be consumed through price and tax incentives, the economic benefits of both of these solutions are far from clear. With regard to ethanol, one can well understand the government's motives for promoting the consumption of fuel hydrous ethanol in particular, given the low international prices of sugar and the current dependence of car manufacturers on this consumption.[25] However, while avoiding more difficult adjustments, this strategy overrates the oil import substitution value of ethanol. It implies government subsidies on production investments, and a reduction of government net revenues, as well as transfers of national income to car owners and ethanol producers, possibly to the detriment of the needs of lower-income groups.[26] Insofar as fuel ethanol is used primarily in private cars, its increased consumption has very limited impacts on overall economic growth. The benefits of exporting surpluses of petrol depend on the real cost of these surpluses, which has not been estimated in this paper nor in Brazil (until at least 1984) due to the lack of detailed data. As we had seen in the discussion in the second section of this chapter, in years in which there is a significant surplus of petrol, the balance-of-payments impact varies considerably under different assumptions about the import cost of exportable surpluses of petrol. This cost needs to be assessed for alternative scenarios of crude oil consumption and crude oil throughput (see note 9), since it determines not only the impact on the balance of payments but also the economic advantage of producing ethanol. Unless such cost is determined, one cannot safely argue whether the fast growth of ethanol output has been economically wasteful or not. Nevertheless, the huge surpluses of both fuel oil (exported at lower prices than crude oil) and petrol, and the fact that there have been deficits only of LPG, suggest that lower crude oil consumption and imports in recent years

would probably have been a more economic option. If so, the cost of exportable surpluses of petrol may be quite considerable depending on how much crude oil consumption could have been reduced economically. And if such cost is high, ethanol output may have grown beyond the economically optimal level.

2. If the net economic benefits of exporting petrol surpluses are low, then it is necessary to increase the oil import substitution effect of ethanol. This will require one or a combination of the following options: (a) investment in the oil-refining sector to reduce the output of petrol and increase that of diesel; (b) decreases in crude oil imports coupled with increases of separate imports of LPG, diesel, and possibly other oil derivatives; and (c) increased substitution of ethanol for diesel through changes in the structure of transport and engines. Changes in the structure of freight transport include various options that would increase the import substitution value of ethanol. While changes from road to rail in freight transport are long-term possibilities and require large sums of capital, the possibility of replacing diesel engines with spark-ignition engines in heavy trucks, buses, and even tractors seems to deserve more attention than it has received so far in Brazil. In addition, a decrease in the gasohol/diesel price ratio could easily promote sales of vans, light trucks, and buses as well as small tractors powered by gasohol or hydrous ethanol. Reducing the growth rate of ethanol output does not seem to be an option preferred by the government. Nevertheless, it should perhaps be considered, particularly if (a) the optimal adjustments in the oil-refining sector continue to yield a high surplus of petrol and the real economic cost of this is close to its export value, and (b) the substitution of fuel ethanol for diesel in utility vans, buses, and lorries remains constrained by economic difficulties in changing the relative prices of diesel, gasohol, and fuel hydrous ethanol.

3. Certainly a lower growth strategy in earlier years would have enabled more systematic planning, which would have been required in order to avoid some unintended distortions (e.g., overconcentration of production in richer areas; concentration of landownership; and crop substitutions that are also associated with a proletarianization of landless farmers who cultivated subsistence crops under rents and other marginal forms of land tenure).

4. Negative effects on income distribution among regions and persons can become increasingly less severe in the future, but this will depend on policies concerning land use, land tenure, and regulation of the sugar-cane supply. The fast growth of ethanol output in 1977–1981, in particular, implied an overconcentration of production in major sugar-producing areas of the richest region and state. As ethanol production increases in poorer regions and states, the effect on inter-regional income distribution will improve, but this geographically-greater dispersion of investment may not necessarily contribute to a more equitable distribution of personal incomes.

Unlike in most southern states, expansion of sugar cane for ethanol in the NE region is more likely to displace crop land than pastures, because many areas with climate and soil quality appropriate for sugar cane also have subsistence crops under cultivation. Much of this cultivation is by small farmers operating under partnerships, rentals, and even illicit occupancies. Furthermore, since these small land plots and illicit forms of land tenure are also much more predominant in the NE than in the C/S regions, substitution of sugar-cane monoculture for food crops may easily cause eviction of peasants and a greater concentration of landownership and incomes than in the C/S regions. CENAL officials are aware of these potential consequences, but to prevent them it would be necessary to change the prevailing criteria for project approval. Both these criteria and the existing financial incentives have certainly been ineffective as a means of avoiding concentration of incomes.

5. The partial findings on the displacement of crops and pastures suggest that the availability of land resources is not a sufficient condition for avoiding crop substitution. In Brazil, this substitution occurred for two major reasons, which were also responsible for the high concentration of investments in the richest state and region. First, the reliance on existing sugar mills implied that annexed distilleries were built in locations predetermined by the geographical and agrarian patterns of sugar production. Second, pressures to reduce raw material costs led to competition for good quality soils with higher yields. In this competition, ethanol farming has had advantages over most food-oriented farming, because of higher investment subsidies, higher prices, and market stability. The lack of effective measures to prevent crop substitutions has been evident. Furthermore, land-use policies involve difficult and lengthy planning processes that were incompatible with the fast-growth strategy adopted in PNA's early years. Improvements have already been made in this regard through careful studies of land use at the state level and through designation of new areas for ethanol. In addition, CENAL can be increasingly more selective of project proposals. Most of these are now for autonomous distilleries that, unlike those annexed to sugar mills, can be located in areas where crop substitutions may be minor.

6. Given the lower demand for car fuel in poorer states, a greater regional spread of production may not promote significant income benefits for poorer states, unless emphasis is given to production based on a relatively larger number of small-scale distilleries rather than on fewer large-scale ones. In principle this would seem a desirable option, since it would proliferate the number of investments and their economic linkages and therefore decentralize the benefits as well. The feasibility of this depends on other factors, such as the financial aspects of small- versus large-scale production, and the implications of a greater decentralization of gasohol-mixing centers for distribution costs. However, the costs of in-

stallation and operation of gasohol-mixing centers are not significant[27] and savings in distribution costs may be possible in some areas, even in the case of anhydrous ethanol. As regards scale, it is true that small-scale distilleries ("micro," varying up to 5000 lpd, and "mini," from 10,000 to 20,000 lpd) have not been able to compete favorably with typical commercial-scale distilleries in financial terms, but it is also true that this has been partly a consequence of the lack of adequate policy incentives and of technological innovation. For example, unlike other scales, micro-distilleries are allowed to produce ethanol for their proprietors' consumption but do not qualify for subsidized investment loans. Mini-distilleries have been given slightly more favorable financing conditions than those of larger scale, but they cannot compete with the latter because ethanol prices paid by the government are the same irrespective of distillery scale. Moreover, the social attractiveness of small-scale distilleries, and in particular those in the micro category, should not be determined by financial criteria alone, because they have a number of advantages that can make them more attractive than larger units for rural development. These advantages include the possibility of making more productive use of distillery effluent (stillage) for direct irrigation and/or production of biogas. Small-scale distilleries could be strategically located in communities where by-products of stillage and surplus bagasse could be used to meet local needs; this could spread out the benefits of ethanol production and promote rural development. In addition, they can more easily avoid crop displacements as their raw material requirement is much smaller and can be relatively dispersed without significant penalties in transport costs.

7. Policy efforts could aim at improvements in the real income of wage earners, mainly through the creation of more stable employment. The author's research found a significant variation in the relative composition of permanent and seasonal labor among distilleries' plantations, largely in consequence of different labor- and farm-management techniques. In principle, more permanent jobs will reduce the number of persons employed, but additional jobs on a more permanent basis can also be created while increasing financial returns on ethanol-related investments. This can be achieved through the promotion of consortia of sugar cane and sweet sorghum for ethanol production. Research projects by government-backed institutions have concluded that these consortia are technologically feasible both from industrial and agricultural standpoints (both crops require basically the same industrial equipment). Since both crops have supplementary cultivation cycles, their use would create more stable employment through extended cultivation and increased industrial production. Production would increase from 150–180 days to about 300 days per year and, for this reason, it would increase financial returns on investments as well. However, the actual implementation of these objectives depends on (a) the demonstration effect of ongoing research by government-backed in-

stitutions concerning crop rotation and consortia, (b) technical assistance, and (c) credit incentives. In addition, the enforcement of existing legislation requiring the plantations to devote a certain amount of land for workers' subsistence farming would be desirable, insofar as the wages of the vast majority of PNA's workers are extremely low.

8. A more equitable distribution of social benefits would also require policy efforts aimed at increasing the number of relatively small farmers who supply raw material. If this were supplemented by the effective enforcement of the "Sugar-Cane Statute" or similar legislation granting external suppliers a minimal share of supply, the current trend toward increased concentration of landownership on the part of the distilleries and their proprietors could be partly avoided. These policies would increase the unit costs of ethanol to the extent that sugar cane from external suppliers means higher raw material costs and risks to the distilleries. In turn, this would imply higher prices to the car-fuel consumers and/or a decrease of financial returns to the distilleries. Higher car-fuel prices would be more likely to occur, which would decrease consumption, but the effects of this on the economy would be marginal, because of the relatively low importance of cars for economic activity and incomes in Brazil. If gasohol and fuel hydrous ethanol are increasingly consumed by vans, lorries, and buses, it may be feasible to establish different prices for different uses, through distribution of coupons, for example.

APPENDIX

Table 2A.1
Evolution of Sugar and Ethanol Production: Land and Raw Material Requirements and By-Production of Stillage, 1974/75 to 1981/82

Harvest years	Area needed thous. ha	Sugar cane m.t	Sugar m.t	Ethanol m.l	Stillage m.l
1974-75	1,490*	74.5*	6.7	625	7 500
1975-76	1,366*	68.3*	5.9	556	6 672
1976-77	1,756*	87.8*	7.2	664	7 968
1977-78	2,092	104.6	8.3	1,470	17,640
1978-79	2,194	109.7	7.3	2,491	29,892
1979-80	2,340	117.3	6.6	3,396	40,752
1980-81	-	-	8.2	3,786	45,000
1981-82	-	-	-	4,240	51,600

*For sugar production only; ethanol was being produced from molasses.

Source: IAA's official data.

Table 2A.2
Production of Anhydrous and Hydrous Ethanol Per Harvest Year (in m.1.) and Per Calendar Year (in m.bb1)

Harvest and Calendar Years	Anhydrous CY m.bbl	Anhydrous HY m.1	Anhydrous annual incr. %	Hydrous CY m.bbl	Hydrous HY m.1	Hydrous annual incr. %	Total CY m.bbl	Total HY m.1	Total annual incr. %
1975-76		233	-		323	-		556	-
1976	1.7			2.1			3.8		
1976-77		300	29		364	13		664	19
1977	4.6			2.1			6.7		
1977-78		1,177	292		293	-20		1,470	121
1978	10.3			2.2			12.5		
1978-79		2,096	78		395	35		2,491	69
1979	15.1			3.4			18.5		
1979-80		2,713	29		683	73		3,396	36
1980	15.5			7.1			22.6		
1980-81		2,211	-19		1 575	131		3,786	11
1981	11.5			13.7			25.2		
1981-82		1,453	-34		2,787	77		4,240	12
1982	15.7			15.9			31.6		
1982-83		3,550	144		2,273	-18		5,823	37
1983	18.9			24.1			43.0		
1983-84		2,468	-30		5,399	137		7,867	35

Note: Production in the 1980 calendar year (CY), for example, comprises part of the output in 1979/80 and part of that in 1980/81. The above figures for production in calendar years were calculated by taking the averages of two consecutive harvest years (HY). This procedure is required for accounting purposes based on calendar years. One bbl is equivalent to 159 liters.

Source: Elaborated from IAA's official data.

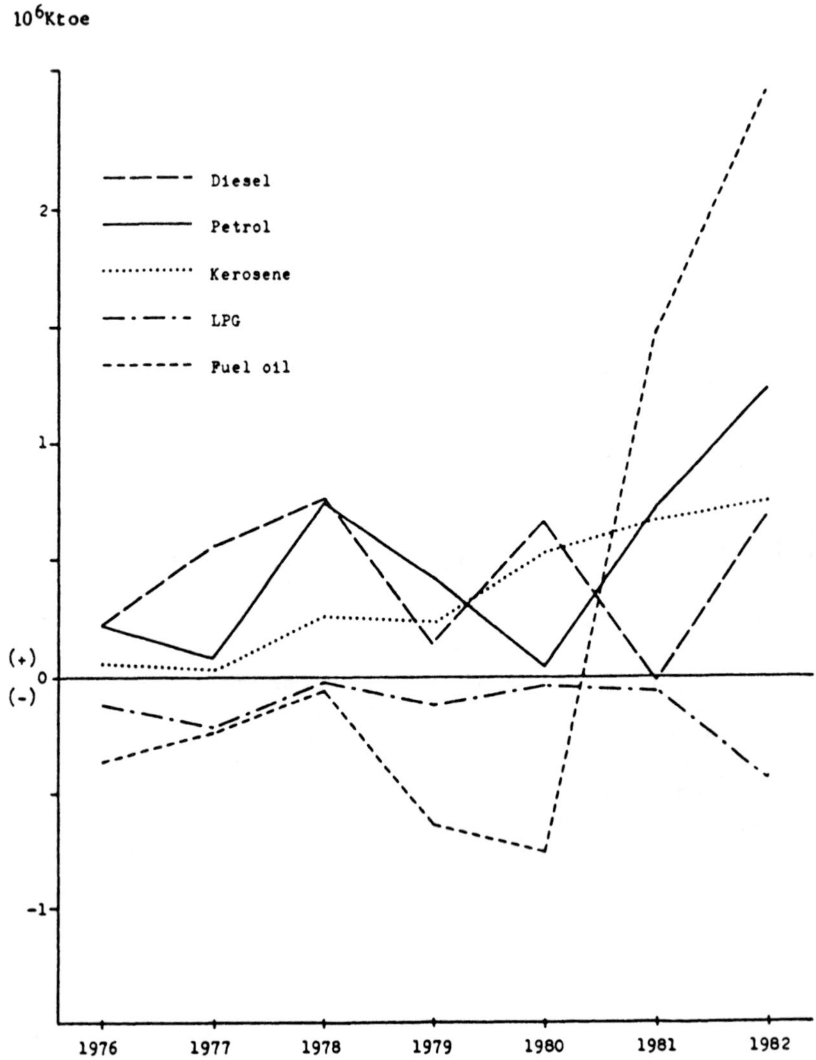

Figure 2A.1
Surplus (+) and Deficit (−) of Oil Products
Source: Elaborated from Brazilian Energy Balance—MME (1983).

NOTES

1. Pereira (1986) includes detailed analyses of capital costs and economic costs of producing anhydrous and hydrous ethanol in different regions and distilleries. The study also compares these costs with economic costs of petrol in terms of equivalent bbl. For example, in May 1980 the economic cost of imported petrol (c.i.f. price plus

distribution cost) was about $57/bbl compared with $51/bbl for anhydrous and $51.7/bbl for hydrous ethanol (also including distribution cost). In May 1981 the figures were $57/bbl for petrol and $52.5/bbl for anhydrous and $54.3/bbl for hydrous ethanol. In 1983, however, the cost of imported petrol may have been between $47/bbl and $51/bbl (depending on a possible variation in the distribution costs measured in dollars) whereas the cost of ethanol would have been no less in dollar terms than in May 1981. Ethanol production costs have also been discussed in COPERSUCAR (1981 and 1983), CENAL (1983), Geller (1984), Ghirardi (1983), O'Keefe (1983), Melo and Pelin (1983), and World Bank (1980b).

2. All of these commercial distilleries were privately owned and used sugar cane and/or molasses, except one owned by the national oil company, PETROBRAS, which used manioc (this unit ceased operations in 1982–83). The most typical distillery has a daily capacity of 120,000 liters, which is normally reached by the third year of operation.

3. These cost figures have been converted from domestic currency into U.S. dollars by using the standard conversion factor, which is the ratio of the value of exports and imports at border prices to their value at domestic prices. For details on this method see Squire and van der Tak (1975), Ray (1984) and Gittinger (1982). The conversion factor thus reflects relative overvaluation or undervaluation of local currency, that is, the difference between the official exchange rate and the parity exchange rate (also referred to as the real or shadow exchange rate); in some periods, it is closer to the "parallel" (black market) rate. The relative value of the local currency is monitored by the Brazilian government on a monthly basis and reported in *Conjunctura Econômica* (various issues; see, for example, Vol. 38, No. 1, Jan. 1984).

An explanation of PNA's investments must be added. A typical autonomous distillery of 120,000 liters per day (lpd) required an investment of about $8.3 million at prices and parity exchange rates of December 1980. This estimate is based on typical overall capital investment of $13 million in the most representative Center/South regions (in Dec. 1980 prices and official exchange rates of Cr$64/$) including distillery, ethanol storage facilities, 5–10 km access road, freight and agricultural equipment, but excluding land. This $13 million is shadow priced at 36 percent due to cruzeiro overvaluation in Dec. 1980—hence $8.3 million. Of this, 80 percent—$6.65 million—was public investment in the form of a loan. In addition, each of the above distilleries required an initial agricultural investment of about $3.3 million—$650/ha (also in December prices and above shadow exchange rate)—including land clearing and preparation, first planting, and initial cultivation. On average, 70 percent of this—$2.3 million—was also subsidized public investment. Therefore, each autonomous 120,000 lpd distillery required an investment of about $11.6 million, including $9 million of public financing. For the production of 1980, 194 of such distilleries would have been required, hence a total investment of $2250 million, including $1750 million of public funds. However, the 207 distilleries in commercial operation included different scales—higher scales costing less, smaller scales costing more, on a per liter basis. Furthermore, 85 percent of production in 1980 was based on annexed distilleries, which required up to 20 percent lower industrial investments than autonomous ones. Furthermore, the 25 percent share of molasses-ethanol in total ethanol output in 1980 implied roughly 25 percent lower agricultural investments than those estimated on the basis of autonomous distilleries.

Taking these factors into account, the above investment figures would in fact have been about $1800 million in total, and $1400 million in public, investments, excluding investments in R & D, gasohol-mixing centers, and other private and public investments in capital goods infrastructure, agricultural R & D, and government services. Including these investments, the final figures may be estimated at $2000 million and $1500, respectively.

4. This conclusion is based on the author's survey (see note 14) and official data from PLANALSUCAR; in this analysis, distilleries' lands include those owned by distilleries' shareholders.

5. Ethanol-related substitutions of sugar cane for crops and pastures has been poorly documented in Brazil. Reliable information available in 1981 concerned four out of six agricultural districts in the state of Sao Paulo (responsible for about 65 percent of Brazil's ethanol output in 1980), and suggested that 18.4 percent of the sugar cane used for ethanol in this state had substituted for pastrues and 10 percent of it for food crops.

6. Brazil is a dualist economy comprising two major macro-regions—the richer C/S and the poorer N/NE. The C/S is subdivided into three regions (South, South-East and Center-West), while the N/NE comprises the North and the North-East, including northwestern parts of the Amazon. This macro-regional classification is not arbitrary and it is used by the government as a basis for national development planning because of the huge disparity of income between the two macro-regions. For example, PNA's financing conditions are different for these two macro-regions. The South-East and South are the richest regions, comprising seven states that are the richest in the country in absolute and per capita terms. The Center-West is poorer in absolute terms and in development infrastructure, but because its population is very low and its per capita income relatively high, it is considered part of the C/S. In the N/NE, the North and most of the Amazon region are poor and thinly populated while the North-East is heavily populated and the poorest region in the country on a per-capita basis. Over 95 percent of the production of ethanol in 1980-81 was in the richest region of the C/S and the poorest region of the N/NE, that is the South-East and the North-East (NE). In 1970, the NE had a per capita income of about one-half the national average and one-eighth of that in the state of Sao Paulo, whereas the per capita of the rural NE was about one-quarter of the national average.

7. This consumption has been promoted through tax incentives to hydrous-ethanol cars and reduced prices of hydrous ethanol.

8. The c.i.f. prices of inputs (imported at the margin) are (in $ millions) 77.8 of diesel, 57 of fertilizers, 27.8 of insecticides, herbicides, and fungicides, and 7.8 of industrial inputs. Data on diesel inputs are from Silva (1976), and on other inputs from CNPq (1978) and official data from PLANALSUCAR and IAA; input prices were based on UNCTAD (1983). Silva's coefficient of 311.7 1/ha of diesel refers to the C/S macro-region. It takes into account manual cutting and mechanized loading during the harvest; an average distance of 10 km between the distillery and the farm; some transport of laborers by lorries, as practiced in Brazil, and other diesel-consuming operations in cultivation. Hence, 3,500 m.1 dividing by 70 1/t equals 50 m.t of sugar cane, which requires at least 0.93 m.ha of harvested land. This multiplied by 311.7 1/ha gives 286.7 m.1 or 1.8 m.bbl of diesel. This is finally multiplied by $43.2, which is the c.i.f. price of diesel ($41 f.o.b. plus a $2.2 shipping cost), totalling $77.8 million. The fertilizer costs were based on PLANALSUCAR's recommen-

dations for fertilizer consumption per ha (i.e., 99.5 kg (N), 141.2 kg (P), and 118.8 kg (K), totalling 359 kg of fertilizers per ha). Although about 57 percent of fertilizer consumption was produced in Brazil in 1980, it is assumed here that PNA increases the demand for imported fertilizers due to the insufficiency of domestic production to meet total demand; 359 kg × 0.93 m.ha = 333.9 m.kg. It is estimated that the use of stillage may have replaced 10 percent of the overall need for fertilizer. Therefore, 300,000 t × $190/t (1980 c.i.f. price for Brazil's imports) totals $57 million. The import costs of insecticides, herbicides, and fungicides were also based on PLANALSUCAR's using the same accounting method. The final figures implied an import total of 4,510 t multiplied by the average price of $3,700/t (c.i.f.), which equals $27.8 million. Industrial inputs include sulphur (for the production of sulphuric acid), ammonia, diamonium sulphate, caustic soda, and anti-foam, which are normally consumed in amounts of (kg/m^3 of ethanol) 2.4, 10.5, 3.7, 0.3, and 0.1, respectively. Enzymes are imported for the production of ethanol from manioc in amounts of about 5 kg/m^3 of ethanol. The minor enzyme requirements for the output of the only manioc distillery operating in Brazil in 1980 were taken into account. Probable imports of steel used in industrial equipment were not included due to the lack of information, although they would not be significant. The import cost of $7.8 million results from multiplying the amounts of inputs, which presumably had to be imported in 1980, by their c.i.f. prices.

The estimated cost of imported capital is based on the assumption that, given Brazil's scarcity of capital and rising foreign debt, government spending on ethanol investments leads to foreign borrowing at the margin. A total government investment of $1,500 for the 1980 production capacity (see end of note 3) at a typical social discount rate of 12 percent over the productive life of investments (20 years) or at 6 percent over 10 years (ignoring grace periods) leads to an annualized repayment of about $200 million, or $8.8/bbl.

9. It would be necessary to trace oil product balances from 1975 (before ethanol production became considerable) to the present. The analysis should be carried out under two hypotheses: one without ethanol, another with ethanol. Each of these situations would then be disaggregated into various scenarios of crude oil throughput to determine the level of crude oil consumption and oil product imports that would have the lowest economic cost. This type of analysis would not only determine the economic cost of exportable surpluses of petrol, which is critical in the estimation of PNA's balance-of-payments impact, but it would also reflect all the different effects of ethanol production on fuels other than petrol. This analysis cannot be done in this study due to the lack of disaggregated data on oil product balances for the 1975–1979 period.

10. Given Brazil's foreign debt situation, it is assumed that refining additional supplies of petrol with domestic capital leads to borrowing at the margin. Petrol refining costs are roughly estimated at $10/bbl in 1980, based on discussions with experts in Brazil (see also *Agroanalysis*, 1979, p. 3). This implies an annual foreign exchange cost of about $1.3 at borrowing terms assumed in the previous chapter. For 1982, the petrol refining cost is estimated at $8, using the same ratio between crude oil price and refining cost in 1980 (see World Bank, 1980b, p. 38), and the foreign exchange cost of about $1.

11. The documented crop substitutions mentioned in note 5 included mainly rice, maize, and cotton. Coffee, castor oil, peanuts, and oranges have also been dis-

placed, but to a very minor extent. All of these except rice and maize are export crops, which might earn/save more foreign exchange than ethanol under certain market conditions. However, data from IBGE (1983) on land-use trends show increases in land use for all the above crops from 1979 to 1982, except coffee. The decline in coffee output, however, is strictly related to the fall in world prices from 1980 to 1983; coffee output in 1983 increased by about 80 percent over 1982.

12. CENAL (1983) provides a detailed analysis of recent changes in the oil-refining structure, further options, and their capital costs, but it does not assess the extent to which these changes and costs are due to ethanol supply. The study questions the benefits of the petrol surplus and exports, but it does not determine the import cost nor the overall economic cost of this surplus.

13. The ethanol production period takes place during the harvest season. Normally, this is 150 days in the C/S macro-region and up to 180 days in the N/NE. Due to the share of the C/S in total production (84.2 percent in 1980), 150 days are used for the purpose of estimating employment. The off-season period varies slightly among distilleries and plantations, but in most cases there is a period of at least 30 days before the harvest, during which practically no activity takes place due to holidays, and so on. As normal practice in Brazil, each planting of sugar cane yields one harvest and two or three ratoons (cuts) before the next planting.

14. Seventy-two questionnaires were sent to the distilleries, and data were collected from 58 units (28 autonomous and 30 annexed), representing 28 percent of the total number of distilleries in operation in 1980 and early 1981. Twenty distilleries were visited in the C/S state of Sao Paulo (14 annexed and 6 autonomous) and 14 in the N/NE States of Alagoas, Pernambuco, and Paraiba (10 annexed and 4 autonomous units).

15. Roughly 85–90 percent of the sugar-cane cutting was done manually in 1980 and early 1981 in the C/S macro-region. The loading of sugar cane into lorries is done mostly by machines with the assistance of labor, although the machine/labor ratio varies among plantations.

16. The partial use of farm machinery may possibly create some economies of scale, but there is no evidence of this. The labor-intensive harvesting of two ha of land will generally require twice as much labor as 1 ha, all other things being equal. One should note, however, that the number of seasonal jobs in the offseason and harvest season do not represent the number of workers, since the jobs may be comprised of alternative compositions of year-round and seasonal workers.

17. This method has some limitations and requires some explanations. Although not all distilleries in operation in 1980 were of 120,000 lpd scale, this scale was the most representative; furthermore, the minor variation in labor/output coefficients among different scales do not lead to considerable distortions in estimating total employment. Considerable distortions result, however, from using autonomous distilleries and labor coefficients in the C/S. Gauging total employment on the basis of this type of distillery ignores the high share of annexed distilleries in total production and their lower demand for labor. This, on the one hand, leads to an overestimation of total direct employment, particularly since it assumes implicitly that all ethanol is produced from sugar cane. On the other hand, the agricultural labor coefficients for the C/S lead to an underestimation of overall agricultural employment, since the N/NE regions (15.8 percent share in total production in 1980–81) have higher agricultural labor coefficients. As it turns out, the net quantitative effect of

these distortions is insignificant—about 2–4 percent overestimation of total direct employment—when all relevant factors are taken into account. The important factors are these: (a) 82.5 percent of the total production of ethanol was based on distilleries with a 20–30 percent lower demand for labor in the administrative/agro-industrial category; (b) about 9 percent of production took place in N/NE areas with labor coefficients twice and three times higher than in the C/S, respectively in the offseason and the harvest season; (c) about 6.8 percent of total production (the remaining share of the N/NE) had overall coefficients about 1.5 higher than those in the C/S; and (d) 25 percent of production was based on molasses, hence not creating ethanol-related employment in agriculture. However, it is best to ignore this insignificant overestimation because the effect of each of the above factors cannot be accurately disaggregated according to occupational categories and types of employment (permanent and seasonal). In addition, since these factors reflect Brazil's conditions only, the quantification of their effect would reduce the relevance of the data for other countries with limited production of sugar. While ignoring some of the above considerations, some recent studies have overestimated PNA's employment—see, for example, Geller (1984) and COALBRA (1983).

18. For example, the total number of workers employed in year-round agricultural activities associated with ethanol production was 24,746 (32 percent of the 77,333 jobs during the harvest season). Moreover, since these would have also been employed during the off season, there would have been 23,587 seasonal workers in the off season.

19. On this subject see, for example, Peek and Standing (1982), Griffin (1974), and Goodman and Redclift (1977).

20. See Graziano da Silva (1978 and 1980).

21. For more details, see Pereira (1986), Graziano da Silva (1980), and Melo (1985).

22. On this subject see, for example, Peek and Standing (1982), Graziano da Silva (1978 and 1980), and Goodman and Redclift (1977).

23. For a review of the concept of income "in kind", as distinguished from cash income, see, for example, Taylor et al. (1980).

24. In real terms, the minimum wage in Brazil decreased steadily from 1959 to 1974; increased by 3–5 percent from 1974 to 1975, but began to decrease again at least until 1980, when major changes in wage policy were introduced.

25. Fuel hydrous ethanol can only be used by hydrous-ethanol cars, and these cannot run on gasohol (or petrol). Since car manufacturers have switched most of their production to hydrous-ethanol cars, they benefit indirectly from incentives to fuel hydrous ethanol consumption.

26. The amount of government loans for main types of investments were described in note 3. The real subsidy over the life of investments, however, cannot be objectively estimated, since it depends on domestic and external inflation over the whole pay-back period and on borrowing rates of interest. Domestic inflation determines how much the government gets back from its loans in real terms; external inflation and borrowing rates determine how much the government pays to foreign lenders. For notional purposes, Pereira (1986, Chapter 3) estimated the cost of the government subsidies on investments as the difference between assumed payments by the government per bbl to foreign lenders and payments received from private investors. The amounted to about $4 in December 1979, $4.3 in May 1980, and $3.0

in May 1981. Government net revenues (measured as the difference between sales at retail prices and total purchases from distilleries) were roughly three times higher for fuel anhydrous ethanol than for fuel hydrous ethanol, because fuel anhydrous ethanol earns the gasohol price, which is much higher than the price of fuel hydrous ethanol. Even when the 17–18 percent lower efficiency of the latter fuel is taken into account, the government revenues from the sale of this fuel are much lower. Thus, the more the government promotes the consumption of fuel hydrous ethanol (e.g., by increasing the gasohol/hydrous-ethanol price ratio) the lower its revenues.

27. Montes and Gomes (1978) estimated capital and operating costs of about $\$50/m^3$ of gasohol.

3

Socioeconomic Aspects of Ethanol Production in India

Ramesh Bhatia

The possibility of expanding the production of ethanol in India for both fuel and nonfuel purposes was contemplated by Indian policymakers during the late 1970s and early 1980s. This possibility was not analyzed in detail, but the government's view (1979–1983) was that priority should be given to the use of ethanol for the production of chemicals, including petrochemicals and pharmaceuticals.[1] This view is based on the assumption that the use of ethanol as a chemical feedstock is more economical than its use as fuel. This assumption cannot be rejected a priori, but it is simplistic and should be assessed in the light of other relevant factors. Such a view of "priority" may in fact have reflected the bias of a lobby of industries buying ethanol at a subsidized price that may have been considerably lower than the price of alternative oil-derived feedstocks.

The collapse of world oil prices in 1986 has made fuel ethanol generally uncompetitive with petroleum products but, as petroleum prices will be likely to change in the future, it remains important to evaluate the economics of expanding ethanol production in India based on cost parameters of the early 1980s.

An ex-ante evaluation of the socioeconomic implications of a large-scale ethanol program in India should consider the following issues:

- projected availability of molasses and the economics of exporting molasses vis-à-vis its use for producing ethanol;

- the possibilities of increasing the output of sugar cane by shifting acreage to regions that have higher yields and/or by providing incentives to farmers to use higher inputs (irrigation and fertilizers) to increase yields;
- the likely impacts of diverting acreage (from food to sugar cane) on food output, food prices, employment, and income distribution;
- the possibilities and the economics of using additional sugar cane for sugar exports vis-à-vis ethanol production;
- the demand/supply balances of oil products (such as naphtha, petrol, kerosenes, diesels) for which ethanol can be a substitute;
- the economics of using ethanol and naphtha as feedstock for petrochemicals for domestic consumption or exports.

This chapter attempts to study these interrelated issues within the framework of Social Benefit/Cost Analysis (SBCA). The SBCA (for further details, see Bhatia, 1983a) is used to evaluate the overall social costs and benefits of two alternative scenarios of producing additional quantities of 2000 m.l and 3000 m.l of ethanol using sugar cane as feedstock. This evaluation also indicates the type of changes required in pricing policies and subsidy/tax policies for inputs and outputs to make such ethanol programs financially feasible (from the viewpoint of entrepreneurs/farmers).

The chapter is organized as follows. The first section gives background data on sugar and ethanol in India, while the second section outlines two ethanol production scenarios for 1990. The third section presents the results of SBCA of these ethanol scenarios, giving details of the estimation of the shadow price for output, economic cost of the major input (sugar cane), and capital costs. The fourth section speculates on the impacts of these ethanol scenarios on output and prices of foodgrains, on employment, and on income distribution. The last section summarizes the results and conclusions.

BACKGROUND DATA ON ETHANOL AND SUGAR CANE IN INDIA

India is the world's largest producer of sugar cane. In 1982/83, the area under sugar cane (3.36 m.ha) was less than 2 percent of the total cropped area and gave a record crop of 189 m.t. Average yield[2] of sugar cane was around 40 t/ha in the early 1950s, but increased to around 48 t/ha in the early 1970s. The yields in 1981/82 and 1982/83 have been higher, at 56.8 and 56.3 t/ha, respectively. There are considerable regional variations in yield rates, varying from 34.5 t/ha in Bihar to 78.6 in Karnataka and 95 t/ha in Maharashtra.

Of the total output of sugar cane, about 48 percent was used for manufacture of *gur* (jaggery) and *khandsari* (unrefined sugar). The output of white (refined) sugar from over 300 factories in 1982/83 was estimated at 8.23

m.t. The installed capacity of ethanol production from molasses was 905 m.1 in 1981/82 while the output was estimated around 700 m.1. About a quarter of total ethanol produced is used for potable purposes and the remainder is used as feedstock by the chemical and the pharmaceutical industries.

In India, tapioca (cassava or manioc), sweet potatoes, and sorghum are used *directly* as foodstuffs, especially by the poor. Any diversion of these items to industrial use or ethanol production is expected to have a serious impact on availability and prices of these foodstuffs and, hence, is not considered socially desirable. However, depending on a variety of factors to be discussed later, the use of sugar cane for ethanol may affect output of foodgrains mainly *indirectly,* by diverting resources such as land, water, and fertilizers away from the production of foodgrains. Nevertheless, our analysis is confined to ethanol from sugar-cane molasses and sugar cane, since these would be the most suitable feedstocks for ethanol in India.

SCENARIOS FOR ETHANOL PRODUCTION IN INDIA: 1990

In this section an attempt is made to outline scenarios for ethanol production in India for the year 1990. This requires projections of demand for and supply of ethanol from molasses; output of and area under sugar cane required for additional ethanol production; and the demand for and supply of major oil products. It may be noted that some of the assumptions underlying such projections and, hence, the results of the SBCA are subject to change. Since it is difficult to measure many of the relevant parameters accurately, one of the methods to take these uncertainties into account is to perform a sensitivity analysis to changes in values of selected parameters as shown in the next section.

The current supply of ethanol in India is from molasses (from the sugar factories) but additional supplies can come from sugar cane or other crops. We have considered two alternative scenarios for 1990: Case I where output of ethanol from molasses (1000 m.1) is supplemented by an additional quantity of 2000 m.1 of ethanol from sugar cane, giving a total supply of ethanol at 3000 m.1. This quantity is assumed to be consumed in the following manner: 340 m.l for potable purposes, 1160 m.l as feedstock for petrochemicals, and 1500 m.1 as fuel. In Case II, it is assumed that an additional quantity of 3000 m.1 of ethanol will be produced from sugar cane, over and above the quantities produced from molasses (1000 m.1). The consumption pattern of ethanol under Case II (Table 3.1) is as follows: 340 m.1 for potable uses, 1160 m.1 as feedstock for petrochemicals, and 2500 m.1 as fuel. The details of various assumptions made in deriving these estimates are described in the following subsections.

Table 3.1
Demand/Supply Balances for Ethanol (m.1)

Items	1978	1990	
		Case I	Case II
Demand/internal consumption[1]	537	3,000	4,000
potable uses	130	340	340
chemical industry	407	1,160	1,160
fuel	-	1,500	2,500
Supply of ethanol	537	3,000	4,000
from molasses[2]	537	1,000	1,000
from sugar-cane	-	2,000	3,000

[1]For 1990, demand for ethanol as feedstock for chemical industry and potable uses are based on the estimates made in Koide et al., 1982. For fuels, it is estimated that ethanol will be used as substitute for diesel/kerosene (details given in Table 3.2).

[2]Assuming that, on average, output of molasses is 45 per cent of the total output of sugar and 1 ton of molasses yields 250 liters of ethanol. It is further assumed that the entire molasses available from sugar factories will be converted to ethanol. This would still leave molasses available from *khandsari* units for other purposes.

Source: See Notes.

Supply of and Demand for Ethanol

Since output of ethanol from molasses depends on the quantity of molasses produced in sugar mills, it becomes necessary to review the projections for demand for and supply of white sugar. Following Harrison et al. (1981),[3] we have adopted 17 m.t as the estimated demand for sweetener consumption for 1990 as given by the historical trend and low consumer expenditure growth (3.5 percent). The share of sugar in the total sweetener consumption is estimated to increase to 45.8 percent by 1990. These assumptions together give the figures for estimated sugar consumption as 7.8 m.t in 1990 (Appendix Table 3A.1).

Exports of sugar depend on a variety of factors including world prices, domestic production, and the ISA quota. India has been exporting sugar for over two decades, although the quantity exported has fluctuated between 70,000 tons in 1977/78 and 2.3 m.t in 1975/76. During the last decade, the unit value of sugar exports fluctuated between \$128/t and \$611/t. According to the World Bank Commodity Price Forecasts of 1980, the long-term projection of the world price of sugar is \$0.16/lb (\$352/t, 1980 dollars). This projected price, when compared with 1980 Maharashtra production costs[4] of about $320/t, suggests that India's sugar exports can be competitive in the world market if production costs can be kept low.

However, apart from the availability of sugar for export and the competitiveness of its price, India's exports are also constrained by ISA quotas. India had an effective export quota of about 0.675 m.t in 1981. However, because of fluctuations in domestic sugar production, the government had to resort to imports to meet the export quota.[5]

It is difficult to estimate India's export quota for 1990. Besides, in view of fluctuations in world sugar prices, it is also difficult to estimate how much sugar the government will be willing to export in 1990. Industry spokesmen speak of increasing exports to 1.0 m.t in years when there is a sugar surplus in the domestic market. Harrison et al. (1981) project an export potential of 1.7–2.5 m.t in 1990. This potential has been calculated from the perspective of supply, based on estimates of production and consumption of sweeteners. Bearing in mind export demand constraints, an estimate of 1.0 m.t has been used for 1990, which may be conservative. As seen later, however, if exports of sugar are relatively more profitable, and if the quota allows it, these can be easily increased by diverting sugar cane from ethanol to sugar. The balance-of-payments aspects of exporting sugar vis-à-vis producing ethanol are discussed in the third section of this chapter.

Using these estimates of sugar output (8.8 m.t to meet domestic and export demands), molasses output is calculated as 45 percent of the total sugar produced. Assuming one ton of molasses yields 250 liters of ethanol, an estimated 1000 m.1 of ethanol from molasses will be available in 1990 (Figure 3.1).

To our knowledge, there is no systematic study that analyzes the economics of using ethanol, naphtha, natural gas, and refinery gas as feedstocks for petrochemicals.[6] Thus, the estimates given for India in the ESCAP regional study (Koide et al., 1982) are used. According to this study, the ethanol-based industry has estimated the requirements of ethanol for its own use at 1160 m.1 in 1990 (Table 3.1). This estimate is higher than the quantities available from molasses, indicating that these requirements as well as those for potable uses will have to come from sugar-cane-based distilleries.

The demand for ethanol as a substitute for petrol/naphtha/diesel has been estimated on the basis of the following assumptions:

- in 1990, gasohol blends with 10 percent anhydrous ethanol and 90 percent petrol will be used in existing car engines. This gives an estimated 400 m.1 of ethanol to replace 0.29 m.t of petrol (about 10 percent of the projected demand for petrol);
- hydrous ethanol can be used straight (100 percent) in small petrol engines for irrigation[7] on small farms and for running other agricultural machinery. It is assumed that the use of ethanol in petrol engines can be made financially profit-

Figure 3.1
Demand/Supply Balances for Sugar Cane, Sugar, Molasses (in m.t) and Ethanol (in m.1): 1990

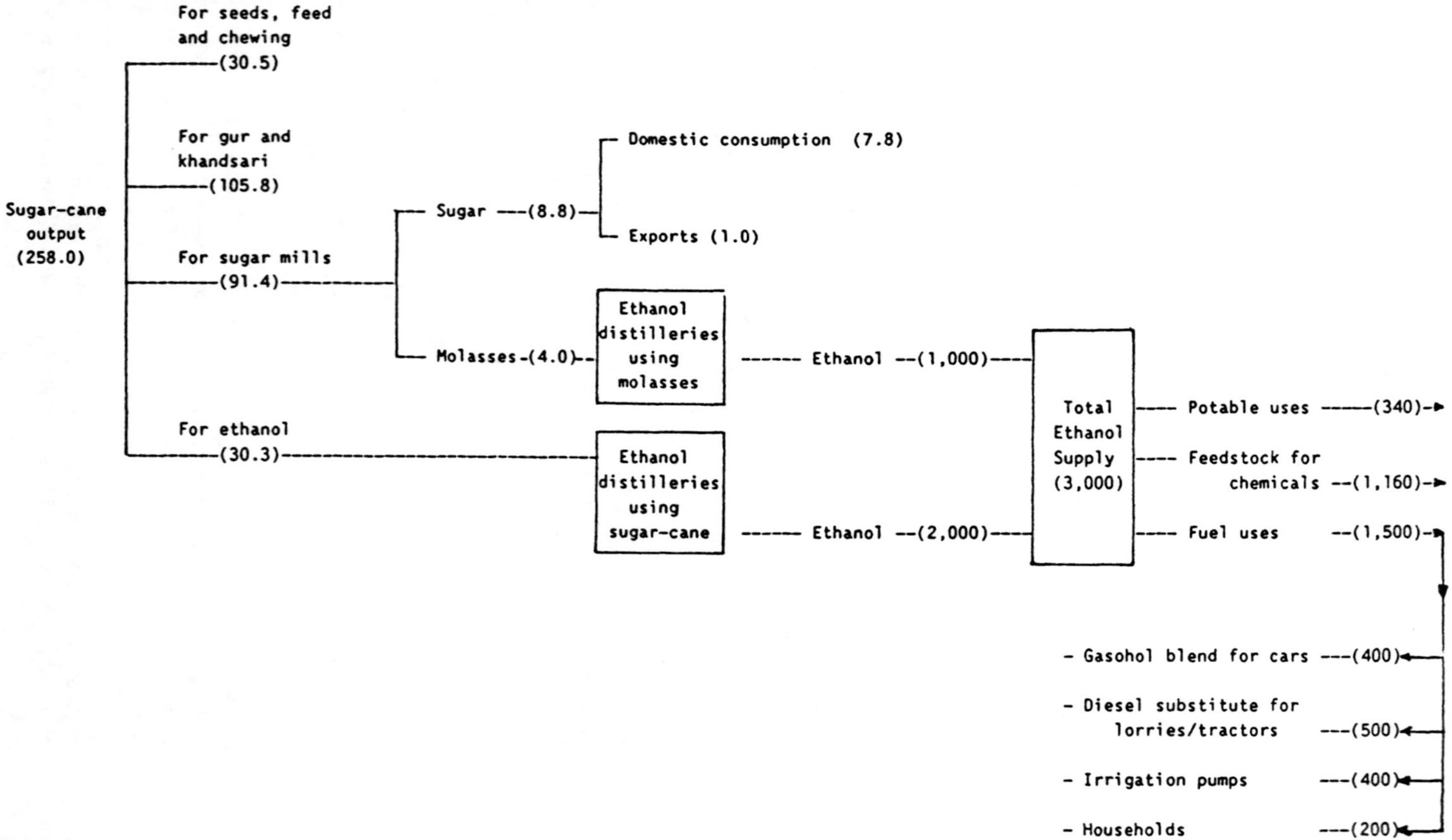

able by making suitable changes in prices of ethanol and petrol;

- blends of diesel and ethanol can be used in stationary diesel engines and vehicles.

In March 1983, there were 2.8 million diesel pumpsets and 4.9 million electric pumpsets in the country for irrigation purposes. Annual increases in recent years have been at the rate of 1.0 million units for diesel engines and 0.4 million for electric motors. There is scope for installation of 12 million pumpsets in the country. We estimate that if petrol engines are not brought into the market on a significant scale, the stock of diesel engines will increase[8] from 2.8 million in 1983 to 5.9 million in 1990. According to available data, diesel engines for irrigation used about 2.5 m.t of diesel oil (including about 1.0 m.t of light diesel oil) in 1982/83. By 1990, such engines may need an additional 2.5 m.t of diesel. In comparison, the total *additional* requirement for diesel in 1990 is estimated at about 11 m.t. We have assumed that about 10 percent of this additional requirement will be replaced by about 400 m.1 of hydrous ethanol used in spark-ignition engines.[9] If a policy of promoting these engines is adopted, the actual substitution may be higher, but we have taken a conservative figure for illustration.

Blends of diesel and ethanol (up to 5 percent) in tractors and other diesel vehicles are technically feasible, but constrained at present by the higher efficiency and relatively lower cost of using petrol in such blends. However, for some remote locations, higher transport and distribution costs may make blends of diesel and ethanol profitable. It is, therefore, reasonably assumed that about 2 percent of the total diesel consumption in vehicles in 1990 may be replaced by ethanol. This gives an estimated requirement of 500 m.1 of anhydrous ethanol to replace about 0.33 m.t of diesel (Table 3.2).[10]

A study by the Institute of Energy Economics, Tokyo, (1981), suggests that applications of ethanol fuel in nonindustrial sectors may include heating and cooking stoves, since there are no technical constraints to their development and use. Of course, use of ethanol as a substitute for kerosene (whose price is subsidized) will require adjustments in prices of fuels and stoves. This is a potential application in India and other developing countries and should therefore be considered. It is assumed that about 5 percent of the additional kerosene consumption in 1990 may be replaced with ethanol. This gives an estimated consumption of 200 m.1 as a kerosene substitute by 1990.

Thus, the total consumption of ethanol for fuel uses[11] is estimated at 1500 m.1 per year in 1990 under Case I (Table 3.2). This consumption (equivalent of 1.06 m.t) is only about 3 percent of the total projected consumption of oil products in that year. This is rather conservative and the actual demand can be higher if ethanol is found socially profitable and made financially viable for consumers. Under Case II, it is assumed that the use of

Table 3.2
Estimates of Ethanol Use as Fuel, 1990

Fuels	Total estimated consumption of oil products (m.t)	Quantity substitutable by ethanol (m.t)		Quantity of ethanol required (m.l)	
		CASE I	CASE II	CASE I	CASE II
Petrol in transport[1]	2.90	0.29	0.29	400	400
Diesel in irrigation[2]	5.00	0.29	0.58	400	800
Diesel in transport (tractors)[3]	18.95	0.33	0.66	500	1,000
Kerosene[4]	8.20	0.15	0.22	200	300
Total	35.05	1.06	1.75	1,500	2,500

Note: See text for further explanations of assumption regarding extent of fuel substitutions mentioned in the notes below. Anhydrous ethanol in gasohol blends substitutes for petrol on a 1:1 basis. For others it substitutes for less. Conversion ratios from liters to tons: petrol, 1355 1/t; kerosene/diesel, 1235 1/t.

1 Assuming gasohol blend of 10 percent ethanol and 90 percent petrol used in existing car engines.

2 Ethanol used in spark-ignition engines for irrigation and other rural uses.

3 Ethanol used as mixture in diesel engines (about 2 percent and 4 percent of total diesel consumption under alternative scenarios, Case I and Case II, respectively).

4 Ethanol used as substitute for kerosene in cooking stoves (5 percent and 10 percent of estimated additional consumption, in Case I and Case II, respectively).

Source: See notes.

ethanol as a diesel substitute (in irrigation and vehicles) could be 100 percent higher than that of Case I, while the use of ethanol as a kerosene substitute could be 50 percent higher than that assumed in Case I.

Demand/Supply Balances for Sugar Cane for 1990

This subsection discusses the demand/supply balances for sugar cane used for the purposes mentioned above. When the demand for ethanol increases as projected in Table 3.1, ethanol molasses is not sufficient and sugar cane has to be used for production of ethanol. Table 3.2 shows that the sugar cane requirement for meeting the estimated demand for sugar (for domestic consumption plus exports) is 78.9 m.t in 1985 and 91.4 m.t in 1990. The

Table 3.3
Requirements of Sugar-Cane Output and Area for Meeting Demands for Sugar, Gur, Khandsari, and Ethanol

Items	Units	1981-82	1990	
			Case I	Case II
Sugar-cane used for sugar[1]	(m.t)	87.2	91.4	91.4
Sugar-cane used for gur and khandsari[2]	"	74.9	105.8	105.8
Sugar-cane used for seeds, feed and chewing	"	21.5	30.5	30.5
Sugar-cane required for ethanol output[3]	"	0.0	30.3	45.5
Total sugar-cane output	"	183.6	258.0	273.5
Sugar-cane yield[4]	(t/ha)	64.4	75.8	75.8
Sugar-cane acreage required	(m.ha)	2.8	3.4	3.6
Sugar-cane acreage as percentage of total gross cropped area	(%)	1.8	2.1	2.2

Note: For details on Case I and Case II of sugar-cane ethanol production, see Table 3.1. The 1981/82 data are from *Indian Sugar* (Nov. 1982). The projections for 1990 are based on the assumptions below.

[1] Sugar output for exports and domestic consumption has been estimated as 8.8 m.t in 1990 (Table 3A.1). Assuming a sugar recovery factor of 9.6 percent, the sugar cane required for sugar is estimated at 91.4 m.t.

[2] The consumption of nonsugar sweeteners is estimated at 9.21 m.t in 1990 by Harrison et al. (1981). Of this, 0.95 m.t is contributed by khandsari requiring 13.3 m.t of sugar cane. The remaining is produced from gur, requiring 92.5 m.t of sugar cane.

[3] 1 ton of sugar cane yields 97 kg of sugar and 12.1 kg of molasses. Assuming that 1 kg of sugar yields 0.65 liters of ethanol and 1 kg of molasses yields 0.25 liters ethanol, the total ethanol output from 1 ton of sugar cane is 66 liters.

[4] Sugar cane yields increased by 25 percent over the decade 1970/71–1971/72 and 1980/81–1981/82. The same rate of increase can be sustained over the next ten years by providing quality seeds, adequate supply of fertilizers (especially in areas with assured irrigation) and by suitable and timely plant protection measures. This gives an estimated yield of 75.8 t/ha for 1990.

Source: See notes.

consumption of nonsugar sweeteners in 1990 is estimated at 9.21 m.t by Harrison et al. (1981). Of this, 0.95 m.t is contributed by khandsari, requiring 13.3 m.t of sugar cane. The remainder is supplied as gur, requiring 92.5 m.t of sugar cane. Thus, the total sugar cane requirement in 1990 is estimated at 105.8 m.t for gur and khandsari[12] and 11.8 percent of total sugar cane production for seeds, feeding, and chewing.

The sugar-cane requirement for ethanol has been estimated as follows: on average, 1 ton of sugar cane yields 97 kg of sugar (9.7 percent recovery) and 12.1 kg of molasses. Assuming that 1 kg of sugar yields 0.65 liter of ethanol and 1 kg of molasses yields 0.25 liter of ethanol, the total ethanol output from 1 ton of sugar cane is 66 liters. Thus, an output of 2000 m.1 of ethanol from sugar cane in 1990 (Case I) will require 30.3 m.t of sugar cane. This gives an estimated total requirement of 258 m.t of sugar cane in 1990, compared with 183.6 m.t in 1981/82 (Table 3.3).

To estimate the area required for producing this quantity of sugar cane, it is first necessary to estimate the average yield in the country in 1990. As discussed earlier, there is significant variation in yields in tropical and subtropical regions. One way of increasing average yield levels would be to shift acreage from the subtropical belt (Uttar Pradesh, Bihar, etc.) to the tropical belt (Maharashtra, Andhra Pradesh, Tamil Nadu, etc.). However there are significant agricultural and socioeconomic reasons why a large part of the area under sugar cane in the subtropics may not be shifted toward the tropical belt.[13] Moreover, in the same climatic regions, yields can be significantly increased by proper selection of sugar-cane varieties and by providing adequate inputs of water, chemical fertilizers, and labor. This can be seen by studying the changes in acreage under sugar cane in a few selected districts of Uttar Pradesh. For example, sugar-cane yields in Saharanput, Bijnor, Meerut, and Muzaffarnagar are almost 50 percent higher than the average yield for the state as a whole. These high-yield districts have a higher proportion of sugar-cane area under irrigation; they also use high inputs of chemical fertilizers. Over a six-year period (from 1970/71 to 1977/78) yields increased by 20–46 percent in the five districts together, accounting for about one-third (6500 ha) of the total sugar-cane area in the State of Uttar Pradesh.

However, use of irrigation and fertilizers would depend crucially on the profitability of sugar cane vis-à-vis other crops. (The private and social profitability of alternative crops is discussed later in this chapter.) Since sugar cane requires bulk transport and sugar recovery depends on how quickly sugar cane can be processed, both area and yields would depend on the certainty of demand for output. If farmers are sure to sell their output to a sugar mill at predetermined prices, then they are more likely to increase acreage under sugar cane. Ray (1981a, p. 14) found that "the inducement to shift area to sugar-cane cultivation in the different districts was primarily transpired by the locational advantages created by the mills in Uttar

Pradesh. The area under sugar cane in each district when correlated with the sugar-cane crushing capacity of the mills located in the corresponding districts provided a correlation coefficient of the order of 0.82." This can also be noted from the micro-level study of 100 farmers from Uttar Pradesh surveyed[14] during 1976/77. Among the five villages selected, two were "gate" villages, that is, those that were near the sugar mill and supplied sugar cane to the gate of the mill. Three were "center" villages, that is, those that supplied sugar cane at selected centers of the Cane Cooperative Union, which in turn transported it to the mill. The farmers in the gate villages supply a relatively higher proportion of their output to the sugar factory and receive higher prices. The farmers in "gate" villages also obtained higher yield and higher net income than those in "center" villages. At a more aggregate level it may be noted that the significant yield increases in Maharashtra (36 percent) and Tamil Nadu (28 percent) over the 1970s are associated with increases in sugar production capacities in these states (Maharashtra 117 percent and Tamil Nadu 98 percent). In contrast, capacity in Uttar Pradesh increased by 26 percent over the decade.[15] These figures show that an assured and guaranteed demand for their output would enable sugar-cane farmers to use higher inputs and attain higher yields.

At the all-India level, the sugar-cane yield increased by 25 percent over the decade 1970/71–1971/72 and 1980/81–1981/82. It is assumed that the same rate of yield increase over the next ten years can be sustained by providing quality seeds, adequate supply of fertilizers, and irrigation.[16] This gives an estimated yield of 75.8 t/ha in 1990.

Assuming an average yield of 75.8 t/ha, the total area under sugar cane required to meet the demand for sugar, gur, khandsari, and ethanol will be 3.4 m.ha. This is about 19 percent more than the area under sugar cane in 1981/82 and is only marginally higher than the area (3.15 m.ha) under sugar cane in 1977/78—the year of the bumper crop. Even this higher acreage under sugar cane is less than 2 percent of the gross cropped area in the early 1980s. Using the same assumptions, 0.4 m.ha will be required to produce 30.3 m.t of sugar cane, which in turn will yield 2000 m.l of ethanol in 1990. For Case II, the quantity of sugar cane required would be 45.5 m.t to produce 3000 m.l of ethanol. The corresponding sugar-cane acreage required would be 3.6 m.ha.

These figures have to be seen in the context of growth of output of sugar cane in the country. As shown in Table 3A.2, output of sugar cane increased by 40 m.t between 1950/51 and 1960/61. The peak output of 189.1 m.t in 1982/83 is higher by 64 m.t compared with the figure for 1971/72. Moreover, the output in the 1980s has fluctuated by 20–30 m.t from year to year. The peak output in 1977/78 was 21 m.t higher than in the preceding or succeeding years. Similarly, output in 1981/82 was 36 m.t higher than in the previous year and 57.6 m.t higher than the lowest production in the 1970s. These figures indicate that for additional output of ethanol, the estimated sugar-cane requirements of 30–45 m.t are not excessively large compared with the current output levels. In fact the estimated requirements are within the range of year-to-year fluctuations in output during the last decade (Table 3 A.2). Although these figures do not justify investment in ethanol production, they do indicate that output and area implications of ethanol programs are not substantial compared with the existing production levels of sugar cane.

SOCIAL BENEFIT/COST ANALYSIS (SBCA) OF THE ETHANOL PROGRAMS

This section presents the results of a SBCA of the ethanol program outlined in the earlier section. According to this methodology,[17] the benefits from ethanol production are estimated in terms of its impact on foreign trade in oil products. Similarly, the cost of ethanol production is estimated in terms of the economic costs of resources (land, water, fertilizer, labor, etc.) that are used for producing sugar cane and ethanol. Thus the analysis is carried out in terms of shadow prices or "accounting prices" rather than market prices of inputs and outputs. The use of shadow prices has become essential in India because of wide distortions in market prices resulting from price controls, taxes and subsidies, import restrictions, and control on foreign exchange and statutory minimum wages. The estimation of benefits and costs at shadow prices is detailed below.

Estimation of Benefits

The estimation of benefits from ethanol requires an estimate of quantities (and prices) of the imported oil products for which ethanol will be used as a substitute by 1990. As discussed in the second section of this chapter, under conservative assumptions of substitution possibilities, about 1.06 m.t of oil products (petrol/diesel and kerosene) will be replaced by ethanol as fuel (Table 3.2). An additional 500 m.l of ethanol (over and above the 660 m.l available from molasses after allocating 340 m.l for potable uses) will be used as feedstock for petrochemicals and pharmaceuticals (Table 3.1).

Table 3.2 gives the estimated quantities of oil products that will be saved when 1500 m.l of ethanol is used as fuel and 500 m.l of ethanol is used as feedstock for the petrochemical industry. The supply/demand pattern of oil products in India is such that any savings in the use of petrol will result in a surplus of petrol or naphtha.[18] In the past, India has not exported petrol, perhaps due to its low octane rating and oligopolistic conditions in the international market. Hence it is assumed that additional output of ethanol as a petrol substitute will result in surplus naphtha for exports or reduction of naphtha imports.

The valuation of naphtha depends on whether naphtha is imported or exported at the margin. Both naphtha imports and exports have been continuing for the last ten years and the quantities imported were higher than those exported. Given the past data, it is difficult to say whether naphtha will be imported or exported at the margin in 1990. However, the spread between import and export prices of naphtha has been marginal and has considerably narrowed in recent years. For example, during 1980/81 and 1981/82, import prices were between \$310/t and \$360/t, whereas export prices were about \$305/t. Moreover, the export price of naphtha has been around 10–14 percent higher than the import price of crude oil in the last four years (1979–1982). In this study, it is assumed that naphtha will be exported at the margin, and its export price (f.o.b) will be of the order of \$300/t (1983 dollars) in 1990.[19] Using this price, the annual benefits from ethanol used as a substitute for petrol and as feedstock for petrochemicals are \$87 million and \$108 million, respectively (Table 3.4).

The remaining quantity of ethanol will substitute for kerosene and diesel, as discussed earlier. It can be assumed that these products will be imported at the margin, since they have been imported continuously during the last two decades and the demand for them is rising faster than that of other products. The import prices of these products have been 30–40 percent higher than the unit costs of crude imports in India. Assuming that the relationship between the prices of middle distillates and crude oil will be maintained, the import price of kerosene/diesel is assumed at \$400/t in 1990. At this price (c.i.f price without premium on foreign exchange) the benefit from ethanol for other uses will be equal to \$308 million per year.

Thus the total benefit from the use of 2000 m.l of ethanol as fuel and feedstock will be \$503 million per year in 1990. In the alternative scenario (Case II), when an additional 1000 m.l of ethanol is used as substitute for kerosene and diesel, the total annual benefit will be \$781 million. Since the entire benefit (export earnings from naphtha and import savings of kerosene/diesel) is in foreign exchange, it must be converted into Indian rupees in order to calculate present values of benefits and costs. As it is difficult to estimate the shadow exchange rate[20] in the year 1990, a notional premium of 25 percent has been used to reflect the scarcity of foreign exchange (Rs 12.5/\$1 instead of the official exchange rate of Rs 10/\$1). Hence, the value of estimated total benefits from ethanol will be Rs 6288 million per year for Case I (2000 m.l of ethanol) and Rs 9763 million per year for Case II (3000 m.l of ethanol).

Besides the savings in foreign exchange, the use of ethanol may also reduce the cost of transporting oil products. This cost reduction, however, depends on the type of ethanol, the location of demand points vis-à-vis refineries/distilleries, and the modes of transport to be used. In view of these difficulties, a notional saving in transport cost is taken as 10 percent of the value of imports/exports of oil products.[21] While needing further study,

Table 3.4

Estimation of Benefits from Ethanol Used as Fuel and Feedstock: 1990 (Case I)

Uses	Ethanol (m.l)	Equivalent Naphtha (m.t)	Naphtha export price ($/t)	Value of exports (m.$)	Equivalent quantity of diesel/kerosene (m.t)	Import price of diesel/kerosene ($/t)	Value of imports (m.$)
Ethanol used as mixture with petrol (10%)	400	0.29	300	87	-	-	-
Ethanol used in petrol engines for irrigation and other uses (as substitute for diesel)	400	-	-	-	0.29	400	116
Ethanol used as mixture in diesel engines (about 2% of estimated diesel consumption)	500	-	-	-	0.33	400	132
Ethanol used as substitute for kerosene in cooking stoves (5% of the estimated additional consumption of kerosene)	200	-	-	-	0.15	400	60
Total ethanol as fuel	1,500	0.29	-	87	0.77	-	308
Additional ethanol used as feedstock in chemical industry	500	0.36	300	108	-	-	-
Total ethanol from sugar-cane	2,000	0.65	-	195	-	-	308

Source: See Table 3.2 and text.

this estimate seems adequate for this preliminary analysis.[22] Thus, the estimated total benefits are Rs 6791 million ($503 × 12.5 + 503) under Case I and Rs 10,544 million under Case II. Since the results of economic analysis will depend crucially on these assumptions, sensitivity analysis to changes in benefits is made later in this section.

Estimation of Costs

The cost of ethanol production can be divided into two major components: (i) capital costs of distilleries and related units, and (ii) operating costs of distilleries including the costs of sugar cane. Costs of sugar cane are expected to account for about 70 percent of the total operating costs; the assumptions made in estimating these costs are detailed in the following paragraphs.

Capital costs of distilleries

No data are available on the capital costs of sugar-cane distilleries in India, since all the existing distilleries are based on molasses. Based on Brazilian data, the World Bank (1980b) has estimated that a sugar-cane-based ethanol distillery of 120,000 lpd (equivalent to about 20 m.l per year, assuming 170 days operation) will cost $7.6 million (in late 1979 prices) in a "low-cost country".[23] India has all the facilities mentioned in that study regarding availability of local equipment and engineering, local construction and implementation, and so on, for similar industries. Hence, it is assumed that India is among the low-cost countries. The total capital costs of 100 distilleries to produce an annual ethanol output of 2000 m.l will be $760 million (in 1979 dollars) or $900–1000 million in 1983 prices.

Since the production of ethanol from sugar cane involves the same basic steps as the production of sugar, it is useful to compare the above cost estimates with the capital costs of modern sugar mills in India. According to available data, a sugar mill in the cooperative sector in Uttar Pradesh that processes 0.18 m.t of sugar cane per year involves a capital cost of Rs 80 million. Based on this, one may estimate that an ethanol distillery processing 0.3 m.t of sugar cane to produce 20 m.l of ethanol will cost about Rs 120 million.[24] This figure is almost 20–30 percent higher than the estimate given in a World Bank report (1980b). This figure is used to estimate the total capital costs of 100 distilleries of Rs 12,000 million, about 60 percent of which is the cost of domestic plant and machinery. Construction of buildings and so forth account for another 20 percent, the remainder being expenses during construction.

Operating costs of distilleries

The World Bank (1980b) does not provide any data on the operating costs of distilleries (excluding sugar cane). Data for Brazil show that the operatings costs of a 20 m.l per year distillery include: (i) $0.19 million for chemicals, (ii) $0.18 million for wages and salaries, (iii) $0.58 million for repairs and parts, and (iv) $0.16 million for insurance. This gives a total of $1.11 million or Rs 11 million per year. Similar information on Indian distilleries is not available. However, data on operating expenses of a sugar mill show these to be Rs 8 million per year for a mill processing 0.18 m.t of sugar cane. Salaries and wages account for about half of these expenses while fuel, power, and consumables account for another 30 percent. These data are used as the basis for estimating operating expenses for labor and other materials.[25] These cost items for a distillery processing 0.3 m.t of sugar cane are proportionately higher (i.e., by 62.5 percent). These operating expenses (at market prices) have been adjusted to arrive at economic costs: by using (i) a notional average shadow price of labor of 75 percent of the market wage rate, and (ii) a premium of 25 percent on the component of fuel oil, power, and consumables.[26] The resulting costs per distillery are Rs 12.50 million at shadow prices. Thus, the total annual operating expenses[27] for 100 distilleries are Rs 520 million for wages and salaries and Rs 750 million for other costs.

Economic costs of sugar-cane production

As discussed earlier, the estimated costs of production of sugar cane (at market prices) were around Rs 100/t ($10/t) in 1978/79 and 1979/80. The total costs of cultivation (excluding the rental value of owned land) increased from Rs 1598/ha in 1967–1969 to Rs 2327/ha in Uttar Pradesh. In the late 1970s the corresponding costs were Rs 5489/ha for Maharashtra, Rs 4547/ha for Andhra Pradesh, and Rs 5814/ha for Tamil Nadu. The conversion of these costs into economic costs in explained below.

Table 3A.4 shows the major components of the cost of cultivating sugar cane across states. In all the states, labor is the predominant component, followed by fertilizers. Irrigation accounts for 7–16 percent of the total costs, except in Bihar. Even though proportions may look similar across states, the levels of various cost components are significantly different across regions. For example, in the subtropical belt, fertilizer costs are Rs 200–260/ha while in the tropical region fertilizer costs vary from Rs 906/ha in Andhra Pradesh to Rs 1223/ha in Maharashtra. Irrigation costs are also substantially higher in the tropical region. Furthermore, the crop rotations vary from one region to another, and it may be necessary to study crop substitution in different states. However, due to the lack of data and difficulties of projecting regional distribution of additional acreage under sugar cane, economic costs are estimated for two regions: subtropical and tropical. It is assumed that this division will take into account the substantial differences across regions that may contribute to additional output.

The first step in the calculation of economic costs is to estimate the opportunity cost of land or economic rent of land at shadow prices. This requires reestimation of the value of output and costs of production of competing crops that will be replaced when sugar cane is cultivated for ethanol production. As discussed in more detail in the fourth section, below, it is difficult to estimate which crops will be substituted when additional area is put under sugar cane. This will depend, among other things, on the relative prices and profitability of different crops, on the scarcity of various inputs that are required in different proportions at different times, and on the demand conditions for output of various crops. Due to these uncertainties, the estimates described below must be taken as preliminary.

Table 3.5
Estimates of Opportunity Cost of Land in the Subtropical Region (Uttar Pradesh and Bihar): (Rs/ha)

Item	At market prices		At shadow prices		
	Paddy	Wheat	Paddy	Wheat	Paddy + 1/2 wheat
Yield (t/ha)	2.33	2.33	2.33	2.33	-
Value of output[1] (including by-products)	2,540	2,900	3,750	5,625	6,560
Input costs[2]					
- Labour	492	459	369	343	540
- Animal and machines	385	584	385	584	677
- Seed	96	160	96	160	176
- Fertilisers	156	520	195	650	520
- Irrigation	350	450	437	562	719
- Other costs	150	116	150	116	208
Subtotal	1,629	2,289	1,632	2,415	2,840
Net profit (before taxes)	911	611	2,118	3,210	3,720

[1]The price of paddy is Rs 1500/t, which is equivalent to a rice price of Rs 2250/t assuming a ratio of rice to paddy of 0.66. It is assumed that the estimated c.i.f price of wheat would be $180/t; which amounts to Rs 2250/t, when given a premium of 25 percent for foreign exchange.

[2]Labor is shadow priced at 75 percent of the market wage. Costs of fertilizers and irrigation are shadow priced at 1.25 times their market prices. Other costs include manure, insecticides, and interest on working capital. It is assumed that fixed capital will be used for sugar cane as well as for competing crops. See text for further explanations.

Source: Author's estimates; see notes and text.

Tables 3.5 and 3.6 detail the assumptions used in estimating the opportunity cost or economic rent of land for the subtropical and tropical regions, respectively. For the subtropical region, it is assumed that a farmer with irrigation facilities has the option of using one of two crop rotations in a two-year period:

i. sugar cane (planted) followed by sugar cane (ratoon) followed by a crop of wheat; or
ii. paddy and wheat in the first year, followed by paddy (or any other *kharif* crop) plus wheat in the second year.

Under these assumptions, a sugar-cane crop will be planted in the month of February/March, continued with a ratoon crop in the following winter that will be harvested some time during November of the next year. With the varieties of wheat crop that can be sown late, the farmer is in a position

Table 3.6
Estimates of Opportunity Cost (Rs/ha) of Land in the Tropical Region

	Market prices			Shadow prices		
Item	Jowar (Maharashtra)	Paddy (Tamil Nadu)	Paddy (Andhra Pradesh)	Jowar (Maharashtra)	Paddy (Tamil Nadu)	Paddy (Andhra Pradesh)
Yield (t/ha)	0.7	3.26	3.01	0.7	3.26	3.01
Value of output[1] (including by-products)	918	3,684	2,868	1,775	5,363	4,885
Input costs[2]						
- Labour	200	794	751	150	596	447
- Animals and machines	162	391	300	162	391	300
- Seed	25	137	103	25	137	103
- Fertiliser	32	415	455	40	519	570
- Irrigation	9	110	158	11	138	198
- Other costs	60	195	200	60	195	200
Subtotal	488	2,042	1,967	448	1,976	1,818
Net profit (before taxes)	430	1,642	901	1,327	3,387	3,067

[1] For jowar, Rs 200 is added as by-product value. The price of paddy is Rs 1500/t and of jowar Rs 2250/t.

[2] See Table 3.5, note 2 for explanations of shadow pricing of major inputs and see text regarding price differences between the tropical and subtropical regions.

Source: Author's estimates; see above notes and text.

to have two sugar-cane crops plus a wheat crop in a two-year cycle. The yield level of the ratoon crop, which is generally lower than the yield from a planted crop, can be significantly raised by proper selection of varieties and application of inputs. Under these assumptions, a farmer has the option to cultivate two sugar-cane crops plus a wheat crop against two paddy crops plus two wheat crops. Thus, at the margin, one sugar-cane crop replaces one paddy crop plus one half wheat crop, and the benefit forgone could be estimated from the net income from these crops. This is shown in Table 3.5 where net income is calculated from paddy and wheat crops using shadow prices for major inputs and outputs.[28] The shadow price of wheat is estimated from an import price of $180/t,[29] which gives a domestic price equivalent to Rs 2250/t, assuming a 25 percent premium on foreign exchange (Rs 12.5/$1 instead of the official exchange rate of Rs 10/$1). It is further assumed that wheat will be imported at the margin and substituted for rice, leading to a price of Rs 1500/t for paddy (assuming a rice/paddy ratio of 0.66). After these adjustments, the value of output at shadow prices is almost 50 percent higher than that at market prices in the case of paddy, and almost 90 percent higher in the case of wheat. The changes made in the cost of cultivation (at market prices) of rice and wheat in the subtropical region to reflect economic costs are:

i. a shadow wage rate at 75 percent of the market rate (as discussed earlier);
ii. a 25 percent increase in fertilizer costs to reflect the foreign exchange premium (since it is imported at the margin); and
iii. a 25 percent notional increase to reflect the possible underpricing of irrigation water.[30]

The resulting estimates of cultivation costs are: Rs 1632/ha for paddy and Rs 2415/ha for wheat (Table 3.5). Thus the net profit from paddy plus one-half of the wheat crop is calculated as Rs 3720/ha. This gives an estimate of the opportunity cost of land used for growing sugar cane in the subtropical region. The economic costs of other resources are added to this to obtain the resource costs of producing sugar cane in this region.

The estimation of economic costs of production of sugar cane in the subtropical region is given in detail in Table 3.7, based on the following assumptions:

i. the current (1983) cost of labor will be 100 percent higher than that estimated for 1979/80 in Uttar Pradesh and a shadow wage rate of 75 percent of the market wage rate is used;
ii. irrigation costs in 1983 are 100 percent higher than the estimated levels in 1979/80 in Uttar Pradesh and a 25 percent notional increase is taken to reflect the underpricing of irrigation water;
iii. the costs of fertilizers are taken as 100 percent more than the 1978–1980 costs in Uttar Pradesh and given a 25 percent premium to reflect shadow prices;
iv. the cost of seeds is also doubled, almost equal to that in Maharashtra.

Table 3.7
Economic Cost of Production of Sugar Cane in the Subtropical and Tropical Regions (Rs/ha)

Item	Subtropical region[1]	Tropical region[1]	
		Subregion I (Maharashtra)	Subregion II (Tamil Nadu and Andhra Pradesh)
Land[2]	3,720	2,654	3,227
Other input costs[3]			
- Labour	1,150	1,238	1,300
- Animals and machines	283	210	112
- Seed	435	435	668
- Fertiliser	673	1,530	1,400
- Irrigation	638	1,120	864
- Other costs	156	102	1,009
Subtotal	3,335	4,635	5,353
Total cultivation cost	7,055	7,289	8,580

[1] The major sugar-cane-growing states in the subtropical region are Uttar Pradesh, Bihar, Punjab, and Haryana. The major states in the tropical region are Maharashtra, Karnataka, Tamil Nadu, and Andhra Pradesh. It is assumed that one half of the area required for ethanol will be from the subtropical region at an average yield of 56 t/ha and the other half from the tropical region at an average yield of 95.6 t/ha.

[2] See Tables 3.5 and 3.6. Rs 3720 is the estimated opportunity cost of land used for growing sugar cane in the subtropical region (p.86). For the tropical region, two estimates are given, one for subregion I and another subregion II. In Maharashtra (subregion I), the opportunity cost of land is estimated as the net benefit (profit) forgone from two crops of jowar (i.e., 2 × 1327 = Rs 2654) as given in Table 3.6. In subregion II, the opportunity cost of land is taken as the average of the net benefit forgone from one crop of paddy in Tamil Nadu and Andhra Pradesh. This gives an estimate of Rs 3227, which is the average of Rs 3387 and Rs 3067 (Table 3.6).

[3] See Table 3.5, note 2 for explanations of shadow pricing of major inputs and see text regarding price differences between the tropical and subtropical regions.

Source: Author's estimates; see above notes and text.

This gives a total cost of Rs 3335/ha, excluding the cost of land. The total cost of cultivation at shadow prices can then be estimated at Rs 7055/ha, including the estimated opportunity cost of land, Rs 3720/ha. It is assumed that with these high levels of input use, the average yields will be equal to or higher than the estimated yield of 56 t/ha in the subtropical region.[31]

Similar calculations have also been made for the tropical region as displayed in Table 3.6. For Maharashtra, *jowar* is considered as the main alternative crop in regions where sugar cane is cultivated. This is mainly due to topography of land where jowar (without irrigation) can be grown in the absence of sugar cane. Since there is an *Adsali* (18 months) crop in parts of Maharashtra and minimum crop duration is 12 months, it is assumed that the opportunity cost of land would be equal to the net profit from two crops of jowar or a combination of jowar and *bajra* crops. The estimated net profit from jowar is Rs 1327 or Rs 2654/ha for land used in growing sugar cane. This is added to the resource costs of producing sugar cane with adjustments made for shadow prices as described earlier in the context of subtropical region. It may be noted that the costs of fertilizers in Maharashtra are more than double the estimated costs for Uttar Pradesh. Similarly, the costs of irrigation are also higher by 80 percent when compared with those for Uttar Pradesh. These figures, when added to the economic cost of land, give the estimated economic cost of cultivation of sugar cane as Rs 7289/ha in Maharashtra (subregion I), a major producer of sugar cane in the tropical region. The corresponding values for the remaining tropical zone are taken as the average of those for Tamil Nadu and Andhra Pradesh (subregion II), assuming that the sugar-cane crop will replace one crop of paddy at the margin.[32] The estimated average cost is Rs 7720/ha if it is further assumed that two-thirds of the additional acreage under sugar cane in the tropical zone will come from Maharashtra (subregion I) and the remaining from subregion II.

All the above information is collated and presented in Table 3.7 for both the subtropical and tropical regions. Annual economic costs of sugar cane for an area of 0.4 m.ha are estimated at Rs 2822 million ($282.2 million) for the subtropical zone and Rs 3088 million for the tropical region. The average costs will be Rs 2950 million per year if the total acreage (0.4 m.ha) is divided equally in the two regions.

This total cost of production of sugar cane must be increased by the cost of transporting sugar cane from fields to sugar mills/distilleries. The transport cost depends on the distances involved and the mode of transport. For example, this cost is probably lower in the subtropical region where carts are pulled by he-buffaloes than in Maharashtra where diesel lorries may be the main mode of transport from fields to factories. It is not possible to estimate transport costs without knowing the locations of distilleries and regional distribution of additional sugar-cane acreage. However, to underline the significance of this component, a notional figure of 10 percent of the total costs of production of sugar cane (i.e., Rs 295 million) has been used as an estimate of the economic cost of transporting sugar cane. This is equivalent to an average transport cost of Rs 10/t over and above the average sugar-cane production cost of Rs 97/t. Although these cost estimates are used in the reference solution, changes in critical assumptions regarding economic costs of sugar cane are also considered.

Based on the above benefits and costs, the results of the economic profitability of the ethanol programs are presented in the next subsection, followed by results from sensitivity analysis.

Results of the SBCA

The results of the SBCA have been presented under two assumptions: (i) unlimited sugar exports, and (ii) sugar exports limited to 1.0 m.t per year (see the second section of this chapter).

Assuming unlimited exports, an additional 2.9 m.t of sugar will be available for exports if not diverted to produce 2000 m.l of ethanol in 1990. As discussed in the previous chapter on Brazil, the international sugar market is very volatile, with large-scale fluctuations in prices that have ranged between \$0.07–0.08 and \$0.40 per pound. A price of \$0.16 per pound or \$350/t in 1990 is used, based on the long-term projection of world sugar prices by the World Bank (1980b). Given the ISA export quotas, India may not be able to export more than about 1 m.t at the estimated price, which may earn \$350 million. The estimated benefit from use of ethanol as a substitute for petrol/diesel is \$503 million, giving a total benefit of \$853 million (Table 3.8). In contrast, the total benefit from exporting 3.9 m.t of sugar at \$350/t is \$1365 million. However it may not be possible to export additional quantities (if permitted under the ISA) at the stipulated price of \$350/t, and prices may have to be reduced to export more. The estimated benefits from sugar exports will be higher than those from sugar exports

Table 3.8
Estimates of Benefits of Producing Ethanol from Sugar Cane vis-à-vis Sugar Export

Item	Assumption I (unlimited sugar exports)	Assumption II (sugar exports limited to 1.0 m.t)
Quantity of sugar exports (m.t)	3.9	1.0
Value of sugar exports (m.\$)[1]	1,365	350
Quantity of ethanol produced (instead of export of 2.9 m.t of sugar) (m.l/yr)	0	2,000
Estimated value of exports of naphtha and imports of kerosene/diesel (m.\$)	0	503
Value of sugar exports plus the value of export of naphtha and import of kerosene/diesel (m.\$)	1,365	853

[1]Sugar export price is assumed at \$350/t (which is the long-term projection made in World Bank, 1980b).

Source: Table 3.4 and text.

plus ethanol production if the average export price of sugar is higher than \$220/t or \$0.10 per pound. Thus the crucial variables that determine the profitability of ethanol programs from the viewpoint of society are the price and quantity of sugar that can be exported by India in 1990 and beyond. This analysis does not depend on the costs of production of sugar cane as long as the export price is high enough to provide incentives for farmers and mills/distilleries to produce the required quantities. If, however, it is assumed that sugar exports will be limited to 1.0 m.t, the question of

Table 3.9
Social Benefit/Cost Analysis of a 2000 m.l/year Ethanol Program in 1990 (Case I)

year	1987	1988	1989	1990	1991	1992	1993 to 2010	Sum of present values
Item		0	1	2	3	4	5	6-22
Present value coefficient at 10% SRD	1.0	0.90909	0.82645	0.75131	0.68301	0.62092	4.09244	-
<u>Benefits</u> (m.Rs)								
Annual benefits from ethanol	0	0	3,375	6,791	6,791	6,791	6,791	-
Present value of benefits P(B)	0	0	2,789	5,102	4,638	4,217	27,792	44,538
<u>Costs</u> (m.Rs)								
Capital costs	4,800	7,200	-	-	-	-	-	-
Present value of capital costs P(I)	4,800	6,545	-	-	-	-	-	11,345
<u>Annual operating costs</u> (at shadow prices)								
Economic costs of sugarcane production and transport	-	-	1,628	3,245	3,245	3,245	3,245	-
Wages and salaries	-	-	520	520	520	520	520	-
Other costs	-	-	730	730	730	730	730	-
Subtotal	-	-	2,878	4,495	4,495	4,495	4,495	-
Present value of operating costs P(C)	-	-	2,379	3,377	3,070	2,791	18,396	30,012

Net present value (m.Rs) $P(B) - P(I) - P(C) = 44{,}538 - 11{,}345 - 300{,}012 = 3{,}181.$

Benefit-cost ratio $\frac{P(3) - P(C)}{P(I)} = \frac{44{,}538 - 30{,}012}{11{,}345} = 1.28$

Source: Elaborated from data in Tables 3.4–8.

economic profitability will require a comparison between the benefits from ethanol and the economic costs of ethanol production. The results of this analysis are presented in the following paragraphs.

Table 3.9 presents the details of estimated benefits and costs of a 2000 m.l program in 1990 (Case I). It is assumed that the program will require a construction period of two years and the output level in the first year of operation will be 50 percent of full production level. It is further assumed that the plant and equipment will have a life of 20 years and about 40 percent of the total capital costs will be incurred in the first year of construction.

Assuming a social rate of discount (SDR) of 10 percent,[33] the present values of benefits, capital costs, and operating costs have been calculated. According to the reference solution using the parameter values discussed in an earlier section, the net present value (NPV) of investment in the ethanol program is Rs 3181 million. The corresponding benefit/cost (B/C) ratio is 1.28. These results show that converting sugar cane to ethanol is economical (from the viewpoint of society) under the assumptions outlined above. Since the estimation of benefits and costs is subject to speculation about foreign trade, production of sugar cane and substitution of ethanol for oil products, it is necessary to add sensitivity analysis to changes in the values of several parameters. This is attempted in the next section.

The results of a more ambitious ethanol program (Case II) of 3000 m.l per year show that the NPV in this case is Rs 9146 million, showing that the present value of benefits (Rs 69,172 million) is higher than the sum of the present values of capital cost (Rs 16,109 million) and operating costs (Rs 43,917 million). The corresponding B/C ratio is 1.57. The 3000 m.l per year program has a relatively higher NPV and B/C ratio compared with the 2000 m.l program;[34] annual benefits are 55 percent higher, because additional output of ethanol will replace kerosene/diesel, which have higher import prices than the export price of naphtha. Capital costs increase by 42 percent, reflecting some economies of scale. In addition, operating costs increase by 46 percent, showing economies of scale in labor costs even though the cost of sugar cane increases proportionately (by 50 percent).

Results of Sensitivity Analysis

This section presents the sensitivity analysis of the above results to changes in (i) economic costs of sugar cane; (ii) capital costs of distilleries; and (iii) estimated benefits from the use of ethanol.

Changes in economic costs of sugar cane

Economic costs of sugar cane may be different from the values used in the reference solution on account of (a) possible variation in the opportunity cost of land (used for sugar cane) in different regions resulting from different assumptions on crop yields, c.i.f prices of food crops replaced, and their cultivation costs; (b) resource costs of cultivation of sugar cane; and (c) lower yields of sugar cane than those assumed in the reference solution. For Case I (2000 m.l per year), Table 3.10 shows the result of changes in the economic costs of sugar cane by considering values that are higher (+10 percent, +20 percent) or lower (−10 percent and −20 percent). If economic costs of sugar cane are 10 percent higher, the NPV decreases to Rs 1051 million. The NPV becomes zero if the economic costs of sugar cane increase by more than 15 percent. This may happen if the c.i.f import cost of wheat increases to about $225/t (from $180 assumed in the reference solution).

The economic costs of sugar cane were calculated assuming that the entire additional area under sugar cane (0.4 m.ha) will displace food crops such as wheat, rice, and jowar. However, if part of the additional area is not under cultivation or is under other less valuable crops/fodder, the economic costs of sugar cane will be lower than those assumed in the reference solution. If so, the social profitability of ethanol production will also be higher. If the sugar cane costs are lower by 20 percent, the NPV increases to Rs 7476 and the B/C ratio increases to 1.66.

The results for Case II show that if the additional output of 1000 m.l of sugar-cane ethanol is mainly used in substitution for kerosene/diesel, the program will be socially profitable even if the economic costs of sugar cane are 20 percent higher than those assumed in the reference solution. At this level, the NPV is Rs 2883 million and will remain positive if the sugar-cane costs do not rise beyond 29 percent of the reference value.

Changes in capital costs

Capital costs of distilleries have been calculated approximately from those of sugar mills. Thus, it is useful to see the impact of variations in capital costs on profitability. For Case I, the NPV is Rs 912 million and the B/C ratio is 1.07 when capital costs are 20 percent higher than those assumed in the reference solution. The NPV will be Rs 5450 million (B/C ratio of 1.60) if the actual costs of these distilleries turn out to be 20 percent lower than those estimated in the reference solution (Rs 12,000 million). Similar results can also be seen for Case II (3000 m.l program) and the NPV remains positive if the actual capital costs are higher by 55 percent as compared to those used in the reference solution. The NPV increases by Rs 2300 million (Case I) when capital costs are lower by 20 percent.[35]

Changes in benefits

Since the estimation of benefits involves assumptions regarding import prices of petroleum products as well as quantities of products replaced by

Table 3.10
Sensitivity Analysis to Changes in Economic Costs of Sugar-Cane Production (Case I, 2000 m.l/year)

Item	Change in economic costs of sugar-cane %				
	-20%	-10%	Reference solution	+10%	+20%
Present value of benefits	44,538	44,538	44,538	44,538	44,538
Present value of capital costs: P(I)	11,345	11,345	11,345	11,345	11,345
Present value of operating costs: P(C)	25,717	27,883	30,012	32,142	34,307
Net present value: P(B) - P(I) - P(C)	7,476	5,310	3,181	1,051	-1,114
Benefit-cost ratio $\frac{P(B) - P(C)}{P(I)}$	1.66	1.47	1.28	1.09	0.90

Source: Elaborated from data in Table 3.9.

Table 3.11
Sensitivity Analysis to Changes in Benefits

Item	Change in benefits (per cent)				
	-20%	-10%	Reference solution	+10%	+20%
Case I (2,000 m.l/yr)					
Net present value (m.Rs) P(B) - P(I) - P(C)	-5,727	-1,273	3,181	7,635	12,089
Benefit cost ratio $\frac{P(B) - P(C)}{P(I)}$	0.50	0.89	1.28	1.67	2.07
Case II (3,000 m.l/yr)					
Net present value (m.Rs)	-4,688	2,229	9,146	16,063	22,980
Benefit-cost ratio	0.71	1.14	1.57	2.0	2.43

Source: Elaborated from data in Table 3.9

ethanol, it is also useful to show the results of sensitivity analysis to changes in level of benefits. Table 3.11 shows that the NPV will increase to Rs 7635 million if the benefits are more than 10 percent higher, compared to those assumed in the reference solution for Case I. However, the NPV becomes negative (Rs 1273 million) if the benefits are 7 percent lower than those estimates in the reference solution. This may happen if the import price of kerosene/diesel is $350/t or the export price of naphtha is $245/t, instead of $400/t and $300/t, respectively.

The above results of sensitivity analysis to changes in parameters show that the economics of both programs are more sensitive to changes in benefits than to changes in economic costs of sugar cane or in capital costs. This clearly suggests that there cannot be a sound policy regarding ethanol production in India based on general notions about the economics of converting sugar cane into ethanol/sugar. Any policy decision should involve careful analysis of the following: prices and possibilities of sugar exports; substitution of ethanol for kerosene/diesel; import/export prices of oil products; the extent and pattern of crop substitution resulting from acreage shifts in favor of sugar cane; and net income from these crops. Under certain circumstances, it may be justifiable to put a premium on domestic substitutes for imported products over and above that on foreign exchange. It may be considered beneficial from the society's viewpoint to produce ethanol, since it may protect the economy from fluctuations in export prices of sugar and import prices of oil products.

IMPACTS OF THE ETHANOL PROGRAMS

Impact on Output of Foodgrains and Other Crops

As discussed earlier, 2000 m.l of sugar-cane ethanol per year (Case I) will require an additional 0.4 m.ha of land, which is approximately 15 percent of the current area under sugar cane (Table 3A.2). As shown in Table 3A.2, this area has increased by about 1.14 m.ha in the last 30 years and 0.21 m.ha in the last decade.

The changes in sugar-cane acreage have occurred in response to an increase in irrigation, an increase in demand for sugar cane by sugar mills, and changes in sugar-cane prices relative to those of competing crops. The analyses of factors affecting sugar-cane acreage have been presented by Narain (1965), Satyanarayana (1967) and Ray (1981a). These and other studies on acreage response functions and relative profitability of sugar cane were summarised in Bhatia (1983). The main conclusions are:

- elasticity coefficients of area response to price are positive and significant and generally less than unity;

- The short-run elasticity coefficient varies between 0.20 to 0.44, depending on the level of aggregation (all India, state/district level), the type of models used,[36] and the type of prices used (this indicates that, on average, a 10 percent increase in the relative price of sugar cane leads to a 2–4 percent increase in sugar-cane acreage);
- the long-run elasticity coefficient is higher, indicating that a 10 percent increase in price leads to a 5–7 percent increase in acreage;
- farmers find it profitable to sell sugar cane to mills that provide guaranteed demand at remunerative prices fixed by the government (because of this, sugar-cane acreage in western Uttar Pradesh increased significantly in districts where new sugar mills were established);
- net income from sugar cane is higher than the sum of net incomes from substitute crops such as paddy, wheat, and jowar.

The above results indicate that the establishment of ethanol distilleries will encourage acreage shifts to sugar cane because farmers will have a guaranteed demand. In order to increase area under sugar cane by 15 percent, the government will have to increase sugar-cane prices by 20–30 percent over a 5–6 year period. High profitability of sugar cane vis-à-vis other crops will guarantee that the required quantities of sugar cane will be forthcoming, as has been the experience in the case of sugar mills. The organizational structure of distilleries will depend on government policy, which is currently to favor sugar mills in the public sector, joint-sector (government plus private), or cooperative sector. Since at least 100 distilleries will be involved, it may be desirable to experiment with alternative organizational schemes in different states.

It is difficult to predict the impact of increased sugar-cane cultivation on crop substitution. This impact will depend on price incentives given to farmers, a number of socioeconomic and agronomic factors, and will vary from one state to another. There is no study on India where such impacts have been quantified in the same way as those reported in Pereira (1986). However the aggregate data for all India, Uttar Pradesh, and selected districts in this state allow one to infer some relationship between area devoted to sugar cane, foodgrains, and other crops. This is discussed in the paragraphs below.

For the period between 1949/50 and 1981/82, sugar-cane acreage increased at a compound growth rate of 2.04 percent per year, while production increased by 3.16 percent. During the same period, area under foodgrains increased by 0.77 percent per year, while output of foodgrains increased by 2.58 percent. Thus, area (and output) of all crops increased along with that of sugar cane, indicating that the increase in sugar-cane acreage was not accompanied by a reduction in area and output of foodgrains. Similar trends can be noted for the period between 1967/68 and 1981/82, when area under sugar cane increased at the rate of 1.92 percent per year.

During this period, area under foodgrains increased by only 0.4 percent per year while their output increased by 2.47 percent, indicating a higher rate of increase in crop yields. Similarly, in Uttar Pradesh, positive growth rates in sugar-cane output were accompanied by positive growth rates in output of all foodgrains and crops. For example, in the western region, a growth rate of 1.45 percent per year in sugar-cane output during the eight year period between 1963–1965 and 1971–1975 was accompanied by a 6 percent annual increase in foodgrain output. These and other similar data show that an increase in sugar-cane output need not result in a reduction of foodgrain output. Changes in irrigation, input use, and adoption of new technologies may be such that these will offset the effects of acreage shifts in favor of sugar cane.

The relevant question is not whether an increase in sugar-cane output results in a decrease in foodgrain output, but whether the additional area used for sugar-cane cultivation would have been used for foodgrains and other crops if it had not been diverted to sugar cane. This question relates to the potential loss of output of foodgrains and other crops rather than the actual reduction that may take place. However it is difficult to estimate what would have been the crop combination and input use in the absence of sugar cane. One may even speculate that the relative profitabilities may be such that increases in foodgrain acreage may follow increases in sugar-cane acreage. The experience of Brazil (Pereira, 1986) suggests that one condition for successful implementation of ethanol programs is to guarantee sugar-cane producers a stable income through fixed but remunerative prices together with a policy of supply quotas. Thus if a small farmer can use part of his land to cultivate sugar cane, this may provide him with the resources needed to use additional inputs on the remaining land under foodgrains. Since foodgrains are required for self-consumption and there are support prices for them, an increase in sugar-cane acreage instead of affecting area under foodgrains may reduce the area used for other commercial crops such as oilseeds, cotton, jute, and tobacco. Some evidence of the second proposition is available for selected districts in Uttar Pradesh, where the area under sugar cane increased during the period 1963–1965 and between 1971–1973. Foodgrain acreage increased in all districts of Uttar Pradesh where sugar-cane acreage had gone up. However, area under oilseeds declined in Basti, Ghazipur, Meerut, and Moradabad. Area under cotton and tobacco declined in all districts except in Basti. Area under jute declined in Basti, Ghazipur, Agra, Moradabad, and Saharanpur.

The only empirical findings that can be used to show the extent and pattern of crop substitution are based on a sample of 40 farmers in villages in the Muzaffarnagar district, Uttar Pradesh. The results show that increase in sugar-cane acreage reduced the area devoted to fodder, wheat, and maize. Reductions in area under foodgrains and fodder were shared almost equally. Among food crops, wheat had the largest reduction in area. The reduction in area under fodder occurred mainly in large farms while medium farms substituted sugar cane for foodgrains. On the whole, not more than 50 percent of the sugar-cane acreage was substituted with foodgrains.

In view of the complex nature of the issues involved, there is a need to carry out detailed empirical studies on crop substitution, foodgrain output, and prices in all major sugar-cane-producing districts, relying on both primary and secondary data. In the absence of any empirical findings, crop substitutions are estimated on the basis of the same assumptions used for estimating economic costs of sugar-cane cultivation in the third section of this chapter. An additional sugar-cane acreage of 0.2 m.ha in Uttar Pradesh may reduce foodgrain output by 0.312 m.t of rice plus 0.232 m.t of wheat (i.e., a total of 0.544 m.t of foodgrains, assuming yields of 1.56 t/ha for rice and 2.33 t/ha for wheat). Similarly, in Maharashtra, an additional sugar-cane acreage of 0.15 m.ha may reduce output of jowar/bajra by 0.21 m.t, assuming an output of 1.4 t/ha for these two crops. The reduction in foodgrain output from the remaining acreage of 0.5 m.ha in Tamilhadu/Andhra Pradesh may be 0.104 m.t (further details in Bhatia, 1983a). The total reduction may be about 0.858 m.t of foodgrains, which is insignificant compared with the total foodgrain output—varying between 109 and 150 m.t in the last five years—or with the year-to-year fluctuations of 5–22 m.t. Even for Uttar Pradesh, a reduction in foodgrain output of 0.544 m.t is within the range of fluctuations normally seen from one year to another on account of weather factors. The possible reduction in food output is also not very large when compared with the possible level of foodgrain imports, which may be 4–5 m.t in years of marginal deficits (of course these can be much higher in years of severe drought conditions).

Impact on Employment

The employment implications of the ethanol programs discussed include direct employment in distilleries, transport, and cultivation of sugar cane vis-à-vis other crops, and indirect employment in input manufacturing, R & D, and government services. In addition, jobs are also created indirectly as a result of increased incomes of farmers and workers. However it is only possible to provide some baseline estimates for direct employment because of lack of data.

Employment in distilleries

Data on employment in ethanol distilleries using molasses in India were not available. As discussed in Chapter 2, this employment cannot be easily estimated because labor used in sugar production is partly used in ethanol production in most cases. Moreover, employment in sugar-cane distilleries is considerably higher than in molasses distilleries.

The alternatives for estimating employment in distilleries using sugar cane is either to use the data for Brazil as a base or to use the available data for sugar mills in India. Since the labor markets are very different in these two countries, it is preferable to adopt the second method, but with the following qualification. Although the overall operation of a sugar mill and a sugar-cane distillery appear quite similar, the employment/output coefficients may be different as a result of different compositions of capital stock and other factors.

The Annual Survey of Industries is India provides information on the number of persons employed by sugar mills and on wages and salaries. In 1973/74, when a total quantity of 38 m.t of sugar cane was processed in 220 mills, the total number of persons employed was 133,710, of which 72 percent were unskilled workers. By 1978/79, the number of persons employed had increased to 313,981, although the quantity of sugar cane processed in 323 mills had increased to 57 m.t (by about 50 percent). It is further estimated that about 65 percent of the employees were seasonal, while the remaining 35 percent were nonseasonal.[37]

If one assumes that sugar-cane ethanol distilleries will provide the same level of employment as sugar mill, 100 distilleries processing 30.3 m.t of sugar cane to produce 2000 m.l of ethanol per year may provide employment to about 800,000 persons per year, using the more conservative data for 1973/74.[38] Since it is not defined whether "nonseasonal" workers (in the offseason) are year-round or seasonal employees, employment may be equivalent to 40,000–54,000 worker-years—averaged to 47,000 worker-years. For the 3000 m.l program, this employment will be higher but by less than 50 percent, since most of the additional output is likely to come from increments of scale in distilleries established for the 2000 m.l program, rather than from new distilleries. One may roughly estimate that employment in distilleries will be 65,000 worker-years.

If correct, these estimates imply that distilleries in India will create about four times more employment than in Brazil. This could be on account of: differences in production techniques and wage rates,[39] and the tendency toward employing more people than are necessary in India, due to social objectives of employment and income distribution.

Employment in sugar-cane transport

Employment generated in transport depends on the type of vehicles used in different regions and the distance travelled from field to factories. For a 2000 m.l program processing 30 m.t of sugar cane, it is assumed that about half of this quantity in the states of North India will be transported by carts pulled by he-buffaloes or bullocks, while the other half will be transported by trucks. It is also assumed that, on average, a cart transports 20 tons per month (five trips per month, carrying 4 tons per trip),[40] and 120 tons during the six months of the harvest season. Thus 15 m.t of sugar cane will require 125,000 carts employing 125,000 persons during this season: 750,000 worker-months or 62,500 worker-years.

Although a part of this employment could be generated by transporting other agricultural output (which would be grown in place of sugar cane), there is a substantial increase in employment when sugar cane is transported because the quantities involved are 20–25 times greater (50–55 tons of sugar cane against 2–3 tons of foodgrains).

Lorries, however, provide much less employment for the same quantity of sugar cane. It is assumed that each truck, with two persons, can move about five times the quantity moved by each cart with one person. Thus the trucks required to move 15 m.t of sugar cane might create 25,000 worker-years.[41] The total employment from transporting sugar cane to the distilleries may therefore be about 87,500 worker-years.

These estimates are not at all comparable to transport-related employment in Brazil, where it is very minor and included in "agro-industrial" and "agricultural employment" (Chapter 2). In Brazil, only lorries are used and, according to Pereira (1986), lorries with a capacity of about 15 tons can make up to 10 trips per day under typical conditions (i.e., most sugar cane supplied from lands surrounding the distillery and loaded mechanically into lorries). The above estimates for India are based on traditional practice, and cannot be interpreted in the sense of "productive employment"; they simply reflect "income employment" given the high surplus and underutilization of labor.

Employment in sugar-cane cultivation

As seen in the previous chapter, the employment impact of sugar-cane cultivation depends on comparative absorption of labor by sugar cane and competing crops. For India, another important consideration is the proportion of family labor in total labor use, since a large part of overall agricultural output is derived from small family plots. Furthermore, this proportion is much higher for sugar cane than for other crops. Although the latest available data on agricultural labor are for the late 1960s, no significant changes in the number of worker-days employed for different crops are expected. Using these data, one can say that, in a two-year period, the total hired labor used on sugar cane (planted) plus sugar cane (ratoon) plus wheat is 118 worker-days/ha. For the same two-year period, two crops of paddy plus two wheat crops require 136 worker-days of hired labor per ha. For total labor, however, the figures are 305 and 258, respectively.

Since the proportion of family labor to total labor depends on farm size, size of family, and so on, it is not possible to project whether the ratios of family to hired labor for sugar cane or other crops will remain stable over time. However it may be reasonable to assume that the relationship of total labor use across different crops will remain more or less the same over time. If so, a shift to sugar cane may result in additional labor requirements of 25 worker-days/ha/year in Uttar Pradesh and other subtropical regions. This means an increase of 5 million worker-days for an additional sugar-cane acreage of 0.2 m.ha (half of the total area).

For the tropical zone, data on Mahrashtra show that total labor requirements for sugar cane are 180–192 worker-days/ha, while those for jowar plus bajra are 122 worker-days—a difference of about 60 worker-days/ha. This means an increase of 12 million worker-days for an additional acreage of 0.2 m.ha in the tropical zone.

Another way of looking at this is to compare labor costs of sugar cane and competing crops. Table 3A.4 gives labor costs of various crops. In the subtropical zone, labor costs for paddy plus "one-half wheat" are Rs 722/ha, whereas those for sugar cane are Rs 765/ha. Against this, labor costs for obtaining estimated yields of sugar cane are Rs 1150/ha—an increase of Rs 430/ha (Table 3.7).[42] Assuming an average wage rate of Rs 10/day, this gives an additional employment of 43 worker-days/ha/year or 8.6 million worker-days for 0.2 m.ha under sugar cane. Labor costs are relatively low for jowar, indicating that sugar-cane cultivation may require labor that will cost Rs 800 more, or implying an additional 60 to 70 worker-days. The difference in labor costs of sugar cane and paddy in Tamil Nadu and Andhra Pradesh are about Rs 430/ha or equivalent to 35–40 worker-days. Thus the additional employment in the tropical zone may be around 40 worker-days/ha or 8 million worker-days for 0.2 m.ha. Alternative methods of estimating additional employment resulting from 0.4 m.ha under sugar cane are very close: 17 million worker-days as compared with 16.6 million worker-days per year. However, it is difficult to estimate the number of jobs that may be created because the additional employment may be in terms of extra work-days for family members and permanent workers already employed on the farm. In this case, additional employment will result in higher income (in cash, or "in kind," or in terms of additional food) rather than in providing extra jobs.

In summary, the estimated direct employment impact of a 2000 m.l program is approximately;

- 47,000 worker-years in producing ethanol in 100 distilleries;
- 87,500 worker-years in transport of sugar cane;
- 17 million worker-days of extra employment in the shift to sugar-cane cultivation (over and above competing crops).

Such a program may therefore provide total additional employment opportunities of approximately 134,500 worker-years in different parts of the country. Since most of these jobs will be seasonal, the number of persons benefiting will be even more—approximately 200,000. In addition, a large number of agricultural laborers would be working for additional days ranging from 25 days per year in the subtropical region to about 40 days per year in the tropical region.

It is difficult to estimate the welfare implications of additional employment in sugar-cane cultivation because this depends on the proportions of hired labor and family labor, the arrangements for paying permanent workers for extra days (e.g., in kind or in food), and the extent to which permanent workers are already gainfully employed and casual labor has to be hired for supplementary work. However, this additional employment is expected to increase the welfare of hired workers, though not necessarily that of family labor, who may have to do extra work without adequate compensation. To the extent that women work as family labor on sugar-cane farms, additional time required may add to their already heavy burden or cut into their leisure time, and result in lower welfare.

Income-Distribution Effects by Farm-Size and Region

The impact of producing ethanol on income distribution will depend on (i) direct and indirect employment created in production, transport, and processing of sugar cane; (ii) relative participation of farmers of different income levels in sugar-cane supply; and (iii) effects of increased sugar-cane cultivation on the pattern of landownership.

On average, total employment (134,500 worker-years) created by the 2000 m.l ethanol program may generate incomes ranging from Rs 600 million to Rs 700 million. These incomes will be received mostly by unskilled and semi-skilled workers in distilleries, poor owners and/or drivers of carts used for transporting sugar cane in North India, drivers of lorries in other states, and agricultural workers. Although cash incomes will be very low in activities such as sugar-cane transport, which will involve high underutilization of labor, there may nevertheless be improvements in incomes. Many of the poor owners and/or drivers of carts do not rely on this activity as the single source of cash income; in fact, some of them will be family members from small sugar-cane farms. A similar argument may be made for a large number of casual workers in sugar-cane cultivation. Many of these will not be employed year-round and, depending on crop rotations in a given micro-region, they may benefit from the additional demand for casual labor. Providing jobs to a large number of people (even though a smaller number could perform the job equally or more efficiently) may be a policy objective so that benefits may be distributed to a larger proportion of the poor in society. Parts of these supplementary incomes, when spent in the rural areas, will lead to additional employment and incomes through a multiplier effect.

The question of income distribution among farmers will depend on the area under sugar-cane cultivation by farm size, and prices received for sugar cane by different categories of farmers. Data on such aspects for different regions are difficult to obtain. As reported in Bhatia (1983a), a microstudy in Uttar Pradesh shows that small farmers rarely allocate smaller portions of their area to sugar cane due to their own self-consumption of foodgrain. However, large farmers sell a relatively higher proportion of their output to sugar mills that, on average, pay prices 15–20 percent higher than those paid by khandsari and gur factories. Thus large farmers located near distilleries will certainly benefit but some small farmers may also receive benefits that were not available earlier. Small farmers are likely to benefit either because of their proximity to distilleries or because of additional demand for sugar cane in the region. Such farmers who are not in a position to sell to sugar mills may find distilleries as buyers. Also those who have to depend on producers of gur and khandsari in years of surplus production may now find another reliable buyer, thus reducing uncertainty and stabilizing incomes. Furthermore, in years of excess sugar-cane production, when high temperatures (especially in North India) lead to a "deterioration" of sugar cane for sugar production, farmers may be able to sell to distilleries since the production of sugar-cane ethanol is not markedly affected by this problem.[43] This may help stabilize the incomes of farmers who receive lower prices due to lower recoveries of sugar on account of delayed harvesting. If the established supply structures of sugar mills obstruct the entry of small farmers, this is not likely to change with the advent of sugar-cane distilleries, and government intervention may be needed to guarantee small farmers a share of supply.

It is also difficult to speculate on the impacts of higher acreage and prices of sugar cane on inter-regional income distribution. As most of the area under sugar cane receives irrigation, regions with higher and better (controlled) irrigation facilities are likely to benefit more. To the extent that irrigation facilities (existing and planned) are unequally distributed among regions (and farm sizes), higher sugar-cane prices and demand will further accentuate these regional (and inter-farm) disparities. However, in the absence of sugar cane, such irrigation facilities utilized for other crops may also accentuate inequalities. On the whole it seems that the income distribution effects may be favorable to lower-income groups in relatively well-endowed regions.

The magnitude of both employment and income-distribution benefits to these groups, however, will depend crucially on policies adopted, particularly on whether the national government is prepared to sacrifice cost efficiency—as suggested by Brazil's experience (Pereira, 1986).

Balance-of-Payments Impact

As shown in the third section of this chapter, the ethanol programs of 2000 m.l and 3000 m.l per year can save/earn about $503 million and $781 million, respectively, through reduction in imports or increases in exports of oil products. These are equivalent to 8.6 percent and 13.4 percent of the total oil import bill of $5,826 million in 1981/82. As the demand for oil products is expected to increase by 80 percent over the next decade, the oil import bill[44] may be about $6–8 million. The direct impact of ethanol programs in 1990 may therefore be 5–8 percent of total oil imports or 4–5 percent of total imports. Indirect outflow of foreign exchange in consequence of ethanol production is not likely to be substantial. There is no reason to presume that equipment will need to be imported and, unlike in Brazil, investment capital is not likely to increase the national debt. The main inputs that will have to be imported at the margin are fertilizers and, to a lesser extent, fuels. Since the total fertilizer consumption will be about 0.1 m.t per year, it will not be a significant proportion of the total ethanol bill. However, a saving of $500–800 million in foreign exchange is substantial. It is difficult to estimate the premium the country may put on this level of savings in 1990, but it may turn out to be crucial at the margin, depending on scarcity of foreign exchange.

Environmental Aspects

The ethanol programs may have environmental impacts through substitution of ethanol for petrol and other oil products as well as pollution effects of ethanol production from molasses and sugar cane.

The high octane rating of ethanol improves the octane rating of petrol in gasohol blends. This eliminates the need to use an octane booster containing lead, which produces adverse effects on the environment. Nevertheless, this environmental benefit from ethanol is of relatively low importance in India and other developing countries where high octane petrol is not generally used. Air pollution is not a hazard in general, although care must be taken to vent carbon dioxide fumes, ethanol vapors, and entrainer vapors (e.g., benzene).

The use of surplus molasses for ethanol production may, in some cases, avoid pollution caused by occasional contamination of drinking water supply. If not adequately treated, distillery effluent—stillage—can cause serious water pollution problems. The capital costs used in this study (World Bank, 1980b) include some facilities for stillage treatment, as typical of Brazil. Furthermore, part of the stillage can be used in direct irrigation and fertilization and can also be reprocessed as mentioned in the introduction to Part I.

Financial Aspects of the Ethanol Programs

Though socially profitable, the ethanol programs will also have to be financially viable to make them acceptable to farmers/entrepreneurs.

For Case I, it is assumed that the market price of diesel/kerosene will increase from Rs 3.25/1 at present to Rs 4/1, giving an output value of Rs 8,000 million per year for the 2000 m.l program. Against this benefit, the costs will consist of (i) annualized capital costs, (ii) annual operating costs, and (iii) costs of sugar cane. Annualized capital charges assuming a 10 percent interest rate and a 20-year plant life (implying a capital recovery factor of 0.11746) amount to Rs 1410 million. Annual operating costs at market prices total Rs 1300 million for 100 distilleries to produce 2000 m.l of ethanol. If it is further assumed that the remaining value of benefits (sales revenue) can be spent on the purchase of sugar cane, this gives a price of Rs 175/t for the purchase of 30.3 m.t of sugar cane. This price is marginally higher than the current minimum price of Rs 155/t fixed by the Agricultural Prices Commission. As noted earlier, the price offered for sugar cane to produce ethanol is lower than the prices fixed by state governments in 1981/82, ranging between Rs 220/t and Rs 230/t.[45] Thus, if ethanol is to be financially viable, the government will have to remove the subsidies on kerosene/diesel and price these products along with ethanol at c.i.f prices corrected for premium on foreign exchange. This will give an ethanol price of Rs 5/1 and a sugar-cane price of Rs 208/t. Since a part of ethanol will be substituted for petrol, which has a considerably higher market price (Rs 6/1, due to high exercise duty), dual prices for ethanol as a substitute for petrol and kerosene/diesel need to be considered. Thus, organizational and pricing aspects of the ethanol program will have to be carefully analyzed, taking into account the interests of all groups—farmers, entrepreneurs and consumers.

SUMMARY OF CONCLUSIONS

Based on the SBCA methodology, the results of this study show that converting sugar cane into ethanol for use as substitutes for oil fuels and petrochemicals is economical (from the viewpoint of society) at oil prices of $300–400/t (1983 dollars) and assuming a limited quota on sugar exports and estimated economic costs of producing sugar cane. As expected, the results are sensitive to changes in economic costs of sugar-cane cultivation and estimated benefits from ethanol. If the economic costs of sugar cane are 15 percent higher than those estimated in the reference solution, ethanol production is not economical. It is found to be economical if benefits are 7 percent lower than those estimated here.

The social profitability of the program considered (2000 m.l and 3000 m.l) are not seriously affected even if capital costs are higher by about 20 percent. In addition, the ethanol programs will have significant impacts on employment and income distribution in several regions. Both programs are much more sensitive to changes in benefits than to changes in the costs of sugar cane and capital costs.

Given the limitations of the data, some of the parameter values used in this study may be overestimates or underestimates and the sensitivity analysis may not be correct for all possible distortions. Nevertheless, this evaluation suggests that the current policy decision of not producing ethanol for fuel uses should be reconsidered, particularly if world oil prices recover to 1979–83 levels. Any coherent policy on ethanol production should not be based on general notions about the economics of converting sugar cane to ethanol/sugar. There is a need for detailed follow-up studies regarding fluctuations in export prices of sugar; fluctuations in sugar-cane output; impact on crop substitution; effects on employment; and the real economic value of foreign exchange earnings from either increased exports or reduced imports of oil products.

Table 3A.1
Demand/Supply Balances for Sugar, Molasses (m.t), and Molasses Ethanol (m.l)

Item	1978-79	1981-82	1990
Demand for sugar[1]	6.2	5.5	7.8
Exports of sugar[2]	0.7	0.5	1.0
Sugar output	5.8	8.4	8.8
Capacity of sugar mills	-	-	9.0
Molasses output	2.6	-	4.0
Molasses ethanol output[3]	537	-	1,000

[1]Data on demand for sugar for 1990 are based on Harrison et al. (1981) assuming that the total sweetener consumption will increase to 17 m.t and will share 45.8 percent of the total.

[2]Sugar exports are estimated at 1.0 m.t for 1985. Harrison et al. (1981) project an export potential of 1.8 to 1.9 m.t for 1985 and 1.7 to 2.5 m.t by 1990. This potential has been calculated from the supply side, based on estimates of production and consumption of sweeteners. However, this does not take into account the demand constraints which depend on the allocation of ISA export quota, etc. Hence, a conservative estimate of 1.0 m.t of exports has been used, bearing in mind that sugar exports can be increased by diverting sugar cane from ethanol to sugar if it is considered profitable at the margin.

The following assumptions have been used in estimating output of ethanol: 1 ton of molasses yields 250 liters of ethanol; output of molasses used for ethanol is 45 percent of the output of sugar.

Source: See notes.

Table 3A.2
Area, Yield, and Output of Sugar Cane

Year	Area (m.ha)	Yield (t/ha)	Production of sugar-cane (m.t)	Sugar cane crushed for white sugar (m/t)	Recovery of sugar (%)	Sugar Production (m.t)	Molasses produced (m.t)
1950-51	1.71	40.5	69.2	11.1	10.0	1.1	0.4
1955-56	1.85	39.3	72.7	19.2	9.8	1.8	0.8
1960-61	2.41	45.7	110.5	31.1	9.7	3.0	1.2
1965-66	2.84	43.7	124.0	36.4	9.7	3.5	1.5
1970-71	2.62	48.3	126.4	38.2	9.8	3.7	1.6
1971-72	2.39	47.5	113.6	31.0	10.0	3.1	1.2
1972-73	2.45	50.9	124.9	40.4	9.6	3.9	1.7
1973-74	2.75	51.2	140.8	42.3	9.3	3.9	1.8
1974-75	2.89	49.9	144.3	48.4	9.9	4.8	2.0
1975-76	2.76	50.9	140.6	41.8	10.2	4.3	1.7
1976-77	2.87	53.6	154.0	49.0	9.9	4.3	2.1
1977-78	3.15	56.2	177.0	67.3	9.6	6.5	3.0
1978-79	3.09	50.2	156.4	59.7	9.8	5.8	2.6
1979-80	2.61	49.4	128.8	39.0	9.9	3.9	1.6
1980-81	3.14	59.4	186.4	87.3	9.7	8.5	3.9
1982-83	3.36	56.3	189.1	82.7	9.9	8.2	3.7

Source: Indian Sugar, Bombay, April 1984.

Table 3A.3
Sugar-Cane Acreage and Production by State—1982-83

States	Area		Production of sugar-cane		Average Yield (t/ha)	Sugar production directly from sugar-cane		Average sugar recovery
	m.ha	%	m.t	%		m.t	%	(%)
Uttar Pradesh	1.80	53.6	82.19	43.5	45.7	2.03	24.7	9.6
Bihar	0.12	3.6	4.18	2.2	34.8	0.37	4.5	8.3
Punjab	0.10	3.0	6.34	3.3	63.4	0.15	1.8	10.6
Haryana	0.15	4.5	5.50	2.9	36.7	0.18	2.2	9.5
Maharashtra	0.33	9.8	31.36	16.6	95.0	3.03	36.8	10.9
Andhra Pradesh	0.17	5.0	12.59	6.7	74.0	0.51	6.2	9.1
Karnataka	0.18	5.3	14.15	7.5	78.6	0.63	7.6	10.2
Tamil Nadu	0.17	5.1	15.21	8.0	89.5	0.65	7.9	9.0
Others	0.34	10.1	17.61	9.3	51.8	0.68	8.1	-
Total (All India)	3.36	100	189.13	100	56.3	8.23	-	-

Source: Indian Sugar, Bombay, April, 1984.

Table 3A.4
Cost Breakdown of Sugar-Cane Cultivation

Cost item	Uttar Pradesh 1979–80		Bihar 1976–77		Maharashtra 1977–78		Andhra Pradesh 1978–79		Tamil Nadu 1975–76	
	cost Rs/ha	% of total cost	cost Rs/ha	% of total cost	cost Rs/ha	% of total cost	cost Rs/ha	% of total cost	cost Rs/ha	% of total cost
Seed	218	6.3	542	13.8	435	6.3	468	7.4	668	7.9
Fertiliser	269	7.8	199	5.1	1,223	17.7	906	14.3	1,120	13.3
Manure + insecticide	71	2.1	128	3.3	101	1.5	75	1.2	428	5.1
Irrigation	256	7.4	64	1.6	1,119	16.2	556	8.7	739	8.8
Labour[1]	765	22.2	647	16.5	1,651	23.9	1,571	24.7	1,733	20.6
Animals and machines	283	8.2	332	8.5	210	3.0	197	3.1	112	1.3
Other charges	1,579	45.9	1,007	51.2	2,182	31.5	2,585	40.7	3,611	42.9
Total cost of cultivation	3,441	100	3,919	100	6,921	100	6,358	100	8,411	100

Note: Data relate to the latest year for which information was available. Sugar-cane yields, in t/ha, were 35.12 in Uttar Pradesh, 42.93 in Bihar, 82.94 in Maharashtra, 62.24 in Andhra Pradesh, and 86.85 in Tamil Nadu.

[1]Labor includes both hired and family labor valued at input cost.

Source: Government of India, Ministry of Agriculture: unpublished "studies on the cost of production of principal crops."

Table 3A.5
Estimates of Net Income and Costs of Sugar Cane and Competing Crops in Uttar Pradesh, 1978/79 and 1979/80 (Rs/ha)*

Item	Sugar-cane (1979-80)	Wheat (1978-79)	Paddy (1979-80)	Paddy plus "one-half wheat"	Paddy plus wheat
Value of main product (cane/wheat/paddy)	6,176	2,421	1,282	-	-
Value of by-products[1]	361	395	212	-	-
Gross income	6,537	2,816	1,494	-	-
Total costs[2]	2,327	1,909	1,751	-	-
Labour cost	765	460	492	722	952
Net income	4,210	907	-257	197	650

*At market prices.
[1]By-products include mainly crop residues.
[2]Excluding the rental value of owned land.

Source: Government of India, Ministry of Agriculture, unpublished "studies on the cost of production of principal crops."

Table 3A.6
Estimates of Net Income and Costs for Sugar Cane and Paddy in Maharashtra, Tamil Nadu and Andhra Pradesh Late 1970s (Rs/ha)*

	Maharastra		Tamil Nadu		Andhra Pradesh	
Item	Sugar-cane (1977-78)	Jowar (1978-79)	Sugar-cane (1975-76)	Paddy (1978-79)	Sugar-cane (1978-79)	Paddy (1978-79)
Value of main product	9,614	719	9,693	3,212	6,782	2,498
Value of by-products[1]	603	206	-	473	100	370
Gross income	10,217	925	9,693	3,685	6,882	2,868
Total costs[2]	5,489	573	5,814	2,352	4,547	2,303
Labour cost	1,651	199	1,733	794	1,571	751
Net income	4,728	352	3,879	1,333	2,335	565

*At market prices.
[1]By-products include mainly crop residues.
[2]Excluding the rental value of owned land.

Source: Government of India, Ministry of Agriculture, unpublished "studies on the cost of production of principal crops."

NOTES

1. Government of India (1977) concluded that "alcohol should preferably be used as a chemical feedstock and not as a fuel, for motive power or a source of energy, because its calorific value per unit of weight is low." Another committee of the government of India (1980) concluded that "in the conditions prevailing in our country, the use of alcohol as a fuel is not considered economic or essential."

2. These are based on data reported by the sugar industry for each year and each state. Since the duration of sugar cane crops varies from 8–10 months to 18 months, a proper comparison of yields would require adjustments for duration. It is not clear from the data whether such corrections were made.

3. The consumption of sweeteners (sugar, *gur*, and *khandsari*) has been estimated by Harrison et al. (1981) in three different ways: (i) using historical data for the period between 1960/61 and 1979/80; (ii) using time-series data for the same period but relating total consumption of sweeteners, total expenditures on all commodities, and real prices; and (iii) assuming two rates of growth of expenditure and no income redistribution. These give a range of 17–20.3 m.t of sweetener consumption for 1989/90.

4. The ex-factory price of sugar in Maharashtra was around $300/t in 1981/82, while the export price was much lower toward the end of 1982. This resulted in sugar export losses, shared by the government and industry.

5. Total world exports of sugar were 26.7 m.t in 1980, 1.8 m.t higher than in 1979. These exports are 30 percent of the total sugar consumption of about 90 m.t. The leading exporters in 1980 were the EEC (4.3 m.t), Cuba (6.2 m.t), Brazil (2.6 m.t), Philippines (1.79 m.t), Australia (2.4 m.t), Thailand (0.5 m.t), and the Dominican Republic (0.79 m.t). The major importers were the United States (3.8 m.t), Japan (2.3 m.t), the EEC (1.4 m.t), and Canada (0.9 m.t).

6. According to the World Bank (1980b), the entire Indian production of acetaldehyde, acetic acid, DDT, a large part of production of acetates, and a small but significant portion of the total production of low density polyethylene (LDPF), polyvinylchloride (PVC), acetone, and styrene have been based on ethanol. A number of naphtha-based petrochemical units have been established and some more have been planned. Discovery of large quantities of natural gas in Bombay High has also made these petrochemicals (based on gas) relatively cheaper. It would be necessary to study these aspects before a firm conclusion on demand estimates can be drawn.

7. In India, petrol prices have been kept very high (through high excise duties) in order to tax the rich who use cars. The price of petrol has been almost twice that of diesel, and petrol engines for irrigation have hence been less economical to users. These engines are very common in other countries (e.g., the United States). If ethanol prices are kept at the level of diesel/kerosene prices, farmers will find it economic to use this multi-fuel engine.

8. The choice between diesel and electric motors has also been distorted in the past due to wrong pricing policies. While diesel was taxed, electricity was sold to agricultural consumers at almost one-fourth (Rs 0.15–0.20/kWh) of the economic cost of generation and transmission. Moreover, the cost of connecting the well to the electricity grid has not been charged to the farmer. Since such large-scale subsidies will not be sustained for a long time and rural electrification costs are becom-

ing high, a shift in favor of diesel/petrol engines may be possible (see Chapter 6 and also Bhatia, 1980a).

9. Use of hydrous ethanol in existing agricultural motors will require mechanical changes. Such costs can be ascertained through field trials and included in analysis at the project stage.

10. The major share of ethanol for diesel substitution will come from new engines. The direct substitution may require additional social costs in engine development and/or imports. Such costs cannot be easily calculated, but should be taken into account in the SBCA. To that extent, the results of this exercise are a bit uncertain.

11. See Introduction to Part I about fuel and nonfuel uses of ethanol.

12. The socioeconomic aspects of production of gur and khandsari versus sugar have not been discussed here since these require primary data and can be a separate research theme. Experts on the sugar industry claim that it is possible to reduce some of the current "waste" of sugar cane used for gur production, since recovery for gur is the same as that for sugar and 1.4 kg of gur is required to substitute for 1 kg of sugar (as a sweetener). Thus, a diversion from gur to sugar increases the availability of sweetener. However, gur and khandsari provide income and employment to a large number of people in the villages of Uttar Pradesh and Bihar. The sociological aspects of gur making also need to be looked into. For some economic aspects, see UNIDO (1980).

13. It is also not clear whether shifting of acreage to the tropical belt is economically necessary, since costs of production of sugar cane in the tropical belt are relatively high, thus giving the same cost of production per ton of sugar cane as in the subtropical region. The opportunity costs of resources used in sugar-cane cultivation will also have to be considered in a detailed analysis.

14. P. Rangaswamy (1978) provides one of the few good micro-level analyses of all the relevant aspects of sugar-cane cultivation. Some of its results are also discussed in later sections.

15. The cases of Uttar Pradesh and Karnataka may, however, require a disaggregated analysis at the district level with adjustments for variations in annual yields.

16. Irrigated area under sugar cane has increased from about 2 m.ha in the early 1970s to 2.5 m.ha in 1978/79. Gross irrigated area has increased from 38 m.ha in 1970/71 to 55 m.ha in 1980/81. Almost 2.5 m.ha of area is likely to receive additional irrigation per year over the next decade. Even if sugar cane maintains its share in the total irrigated area, this should make a substantial irrigated area available for the sugar-cane crop.

17. The methodology of SBCA is well known and has been applied in the case of renewable energy sources/technologies. Some references are Little and Mirrlees (1974), Squire and van der Tak (1975) and Gittinger (1982). For applications in the renewable energy field, see Bhatia (1977,1980a), de Lucia and Bhatia (1980), French (1979) and Barnett et al. (1978).

18. As discussed earlier, higher production of ethanol will be used in gasohol blends for cars in direct substitution for petrol. However, in the refineries it is easy to switch. Reduction of petrol output will be reflected as additional availability of naphtha if there are no petrol exports.

19. If the naphtha export price is assumed to be 12 percent higher than the crude

oil import price of $36/bbl c.i.f, this implies $300/t for naphtha exports.

20. It is outside the scope of this study to estimate the shadow price of foreign exchange. Hence, the premium estimated by the Planning Commission is used, which is 25 percent over the official rate.

21. These transport costs estimated at Rs 503 million (10 percent of Rs 5030 million) have been added to the benefits. This implies that, with the domestic production of ethanol, there will be a reduction in average transport costs.

22. The effect of changing this assumption on the results can be seen under sensitivity analysis to change in benefits (Table 3.11). This and other problems associated with regional costs of production of sugar cane and ethanol suggest that a detailed inter-regional study should be carried out. This is being attempted in the framework of a linear programming model that has been presented in Bhatia (1983a).

23. For details of capital costs in Brazil, see Pereira (1986); see note 3 of Chapter 2 for the main figures.

24. The only data available on a sugar cane distillery of 13.5 m.1 of ethanol per year are from a report by the Ministry of Agriculture. According to this, capital costs are Rs 22.7 million and annual operating costs are Rs 2.8 million.

25. In any case, these expenses account for less than 30 percent of the total costs as the sugar-cane cost dominates the overall cost of production.

26. There are no studies on shadow wage rates for different regions. Using the methodology discussed in Sinha and Bhatia (1983), an approximation has been made. Since the domestic prices of fuel oil, power, and consumables are based on import parity (i.e., c.i.f price), an increase of 25 percent will reflect the premium on foreign exchange. Costs of insecticides and herbicides are included under other costs although their use is marginal in India.

27. No additional energy input is required as noted by Pathak et al. (1980). They find that in an integrated sugar-ethanol complex utilizing modern production methods, by-product bagasse fully meets the thermal, mechanical, and electrical energy requirements of sugar and ethanol production. In Brazil, even the autonomous distilleries are self-sufficient in energy (Pereira, 1986). Thus our figures for costs of fuel and power may be overestimated.

28. For a discussion of methods for estimating shadow prices of agricultural outputs and inputs, see Sinha and Bhatia (1983).

29. A higher import price is considered later in the sensitivity analysis.

30. There are no data to show which type of irrigation is used for sugar cane across regions. In the case of canal irrigation, irrigation charges are not known to reflect opportunity costs. Costs of water obtained from tube-wells are likely to be closer to the economic costs of irrigation. This requires better data than are easily available. See note 26 regarding the shadow wage rate and premium on foreign exchange.

31. Sugar-cane yields in five major sugar-cane producing districts of Uttar Pradesh were higher than 50 t/ha in 1977/78. Consumption of fertilizers (NPK) ranged between 55 and 88 kg/ha. These five districts accounted for 0.65 m.ha of area under sugar cane in 1970/71 but increased to 0.81 m.ha in 1977/78. Thus it is reasonable to assume that about 0.2 m.ha of additional area can be brought under sugar cane in the subtropical region.

32. In Tamil Nadu and Andhra Pradesh, the sugar-cane crop has a duration of 10–12 months. To our knowledge, there is no study of crop rotations in major sugar-cane districts, but it is possible that in some districts another crop may be cultivated in addition to one main paddy crop. This is taken into account in the sensitivity analysis.

33. For details, see Sinha and Bhatia (1983).

34. One cannot conclude from this that the B/C ratio would keep on increasing with additional ethanol production because diseconomies of scale arise when additional ethanol is used for kerosene/diesel.

35. As noted in the third section of this chapter, the actual costs of distilleries may turn out to be less than those assumed in the reference solution.

36. In most studies, variations of the Nerlovian type of model have been used.

37. *Indian sugar yearbook*, 1980/81, p. 161.

38. A total of 133, 710 persons divided by 220 sugar mills equals 607 persons per mill, processing an average of 172,727 tons of sugar cane, which is the amount needed per distillery of 65,000–70,000 1pd. Judging from information in Pereira (1986), a standard 120,000 1pd distillery may require about 30 percent more labor than a 60,000 1pd, hence 800 people per distillery or 80,000 people for 100 distilleries. The available data in India do not permit any reliable estimates of differences in employment levels in sugar mills and sugar-ethanol distilleries. Furthermore, data on Brazil are likely to be very different since some operations that are typically mechanized in Brazil tend to be labor-intensive in sugar mills in India. This is the case for unloading and preparation of sugar cane before the juice extraction stage of production. For a discussion of techniques and employment in different operations of sugar-cane cultivation and ethanol production, see Pereira (1986, Appendices C and D).

39. Data on a new sugar mill processing 0.18 m.t of sugar cane show that Rs 2.9 million consist of wages and salaries: this is similar to data from the Annual Survey of Industries. Wages of unskilled labor in India were estimated to be about one third of those in Brazil (see Pereira, 1986).

40. Although the average distance involved may be 10 km from the field to the sugar mill, a lot of time is wasted in the queues at the sugar mill. Long queues of carts can be seen in the districts of Meerut and Muzaffarnagar, Uttar Pradesh, during the sugar-cane harvest season.

41. It is being assumed that a lorry has a capacity of 10 tons and makes 10 trips in a month (instead of 5 by a cart), thus transporting 100 tons per month. To transport 2.5 m.t per month, the number of lorries required will be 25,000, providing jobs to 50,000 persons over a six month period. This level of employment created will change depending on the topography of land, scale of distilleries, and distances involved.

42. Actual labor costs of sugar cane (Rs 765/ha) in 1979/80 were almost equal to those for paddy plus one-half-wheat (Table 3A.4). However, the average sugar-cane yield in that year was only 35.12 t/ha. It is estimated that 100 percent higher labor inputs (along with other inputs) are required to obtain higher yields.

43. In years of excess sugar-cane production, when the sugar mills operate in May and June, sugar recovery decreases to 7–8 percent compared with the peak

recoveries of 10–12 percent recorded in January–March. In 1977/78, some sugar mills operated in July and August and recovered only 4–5 percent. This is due to inversion of crystallizable sucrose into fructose and glucose. As the latter are not crystallizable, the sugar recovery decreases. But in the case of production of alcohol, the inversion of sucrose into glucose and fructose does not make any difference; in fact, in distilleries, sucrose is converted into glucose and fructose before the fermentation process takes place.

44. Although domestic oil production has increased significantly over the 1980–1984 period, there have been no significant new discoveries. If this situation continues, India will be importing crude oil and products in the early 1990s even if the projected increases in domestic production take place.

45. Even at this lower price, sugar cane will be more profitable to farmers than competing crops.

II

COMPARATIVE MICROECONOMIC EVALUATION OF RENEWABLE ENERGY TECHNOLOGIES FOR IRRIGATION PUMPING

4

Traditional Sources of Energy for Low-Lift Irrigation Devices in Bangladesh and Egypt

David Birch and Janusz Rydzewski

This chapter presents an economic analysis of low-lift irrigation devices using traditional sources of energy: human and animal power. Unit costs of pumping water using traditional energy sources are compared with the costs of mechanized pumps using diesel or petrol engines. Although this study is limited to low-lift irrigation practices in Bangladesh and Egypt, many of the principles discussed and the conclusions drawn may be applicable to other countries with similar conditions. The first section describes the traditional and mechanized low-lift irrigation technologies suitable for Bangladesh and Egypt, and the second section reviews the experience with these devices in these countries. The third section compares the various alternatives in terms of (i) costs per m^3 of water output per season and (ii) present values for a given farm size. The fourth section summarizes the conclusions of this comparison.

Summarized and partly rewritten by R. Bhatia and M. Swaminathan, and edited by A. Pereira. This chapter is a shortened version of D. Birch and J.R. Rydzewski, *Energy options for low-lift irrigation pumps in developing countries: the case of Bangladesh and Egypt* (Geneva, ILO, 1980; mimeographed World Employment Programme working paper, No. 57; restricted).

CHARACTERISTICS OF TRADITIONAL AND MECHANIZED TECHNOLOGIES

Animate energy is still an important means of lifting water for domestic supply and irrigation in many countries. Four types of human-powered lifts suitable for Bangladesh and Egypt are: the *dhone,* the *shaduf,* the *tambour,* (Archimedean spiral), and the Chinese chain and washer lift. The main types of animal-powered lifts are the water wheel or *sakia* and the chain and washer lift. These devices and their water output capacities are described below.

Human-Powered Lifts

The shaduf: This consists of a lever in the form of a pole pivoted near its center like a see-saw. At one end is a bucket made of skin, clay, or a large tin can. At the other end is a counterweight made from either stone or clay. By rocking the pole the bucket is alternately lowered into the water to fill and raised to field level where it empties. The *shaduf* is still used extensively in Egypt and the Sudan and in some areas of Bangladesh, mainly for irrigation of vegetable gardens rather than field crops. It is used mostly for lifting water from dug wells, since the sloping sides of rivers make it difficult to operate. In the Sudan, the average output of a *shaduf* has been estimated as 0.83 1/sec with one man working over a six-hour day and lifting water 2.5 m.

The dhone: This is similar to the *shaduf* except that, instead of the bucket, a hollow log is used (occasionally a modernized version uses a half cylinder made of sheet steel). One end of the log is fixed on the side of the field and acts as a pivot. The other end is lowered into the water and filled. This end is then raised to a level slightly above that of the field; the water then flows down the log. The *dhone* has a low lift, probably 2 m at the maximum, and can discharge 0.9 1/sec (or 19.4 m^3/day) when operated by one man working a six-hour day. The *dhone* is frequently used for irrigating small plots adjacent to rivers in Bangladesh. It has the advantage of moving water across and up the sloping banks of these rivers. Occasionally, two or three are used in series to gain a higher lift. There is very little capital cost involved, labor cost accounting for the bulk of total cost.

Tambour (Archimedean spiral): This is the main human-powered lift used for irrigation in Egypt today. It consists of a cylinder, traditionally made of wood, but nowadays more commonly made of galvanized sheet steel. A spiral is constructed inside the cylinder (in Egypt this is done by mounting strips of wood along the axis of the cylinder, each with an angular displacement from the one beneath it). The cylinder is placed on the bank (usually at an incline of 30°) with the lower end in the water. The top end is fitted with a crank handle to enable one or two men to operate it, though it could be fitted with a gear mechanism to allow it to be driven by a cow. The Egypt Water Use and Management Project estimates that two men (each working a six-hour day) can deliver 15 1/sec at a 1 m head. S. Wilson (1977) notes that an archimedean spiral pump has an efficiency of around 80 percent. Assuming this, with the average head of 1.5 m, two men should be able to deliver 8 1/sec or 173 m^3/day over a six-hour day.

The Chinese chain and washer: This has been known and used for many centuries in China and Europe. It works by the action of the chain washer being pulled in a continuous loop inside a closely fitting pipe, then over a geared wheel and back down to the bottom of the pipe. Water is carried up between each washer from the mouth of the submerged pipe to the top where it is discharged into a trough. The maximum practical lift for this technology is 15–20 m; several chain and washer assemblies can be fitted to the same axle to increase the discharge. It has been estimated that two men can lift 1.7 1/sec or 37 m^3/day over a six-hour day.

Animal-Powered Lifts

The sakia: Traditionally, the *sakia* consists of a vertical wooden wheel with a series of clay pots attached to the periphery. The lower side of the wheel dips into the water source. As the wheel rotates, the pots with water reach the top where they empty into a trough that channels the water to the land. The wheel is driven by one or two animals circling around it. The power is transferred to the wheel by means of a wooden right angle gear drive. The system has been modernized (largely resulting from work carried out at the Hydraulic Research Station in Egypt), with the vertical wooden wheel being replaced by a segmented wheel made from galvanized steel, and the wooden right-angle drive being replaced by a cast-iron gear drive. The sheet-steel vertical wheel picks us water at its periphery, then as the wheel rotates, the water moves in its segment to the center of the wheel where it is discharged at field level. The modernized lift has almost replaced the older type in Egypt, and is very efficient (about 60 percent) in the low head conditions that exist in most areas of the Nile Valley and the Delta. Discharges of around 30 1/sec can be achieved in Egypt from a sakia lifting 1.5 m, and powered by one cow. This would give an output of 1080 m^3/day over a 10-hour day.

Chain and washer: The sakia can lift water only up to about 2 m. For larger heights, the chain and washer is more adequate, provided that it is fitted with a right-angle drive (cast iron) and yoke to make it suitable for an animal drive. This device can lift water from a well or a surface. Watt (1976) gives the output of a chain and washer device driven by two bullocks as 14 1/sec from a height of 4 m (using three chains) and 9 1/sec for a height of 6 m (using two chains). Neither this device nor any other types of animal-powered lifts are widely used in Bangladesh. However, a theoretical cost estimate has been worked out for this type of lift based on the authors' knowledge for animals in the Kastia district. The estimated water output is 324 m^3/day for a 10-hour day.

Mechanized Low-Lift Irrigation Pumps

The most commonly used pumps in this category are those powered either by electricity or by an internal combustion engine using petrol or diesel or, in some cases, kerosene. The demand for these pumps has increased considerably in developing countries, including Bangladesh and Egypt. Both countries import the majority of their pumps and engines, and Bangladesh also has to import oil. Almost all the diesel-driven pumps in Egypt and Bangladesh are of the centrifugal (or mixed flow) type. In these pumps, the pressure head created is largely attributable to centrifugal action. Centrifugal pumps have the advantage that their speed can be matched to that of the driving engine and they can be directly coupled. Medium-speed diesel engines are being used throughout the world as they generally provide an efficient means of turning oil into power on a small scale. Slow-speed diesel engines are much larger and heavier, and have become less popular due to their high capital cost. For smaller sizes, petrol engines can be produced more cheaply. Many manufacturers produce 1–2 kW petrol engines coupled to self-priming centrifugal pumps.

EXPERIENCE WITH TRADITIONAL AND MECHANIZED TECHNOLOGIES

Bangladesh

Irrigation has been practiced in Bangladesh only since the beginning of this century. Since 1960, mechanical pumps have increasingly substituted for human-powered lifts, although in 1979 the latter irrigated about 0.4 m.ha (or 38 percent of the total irrigated area in the dry season).

The two methods of human-powered irrigation used in Bangladesh are the *dhone,* and the chain-and-washer lift. It is small landowners, owning less than 1 ha of land and using their own (or family) labor for irrigation who find such lifts most profitable. During the irrigation season, there is underemployment; thus the opportunity cost of labor for a small farmer is likely to be low.

The costs of operating a *dhone* are thus quite minimal for small farmers (only the low capital cost) compared with large landowners who have to hire labor for the task. The use of simple and locally constructed devices also lowers risks and costs, as machines sometimes break down and spare parts may not be available.

The chain-and-washer lift is more efficient than the *dhone*, but it was not used in 1979/80, possibly because its capital cost is about 18 times higher than that of the dhone or because farmers interested in modern technologies may have been attracted by the diesel-powered pumps, which are subsidized by the BADC (Bangladesh Agricultural Development Corporation). Although animals are not used for irrigation at present, some limited use may be possible in areas with underutilized cattle.

It has been a policy of the Bangladesh Government to increase the use of mechanized pumps in its efforts to raise food production. Mechanized irrigation brings problems with it, however, as large areas have to be irrigated to achieve the economies of scale required to make the operation economically attractive; this is difficult to organize in Bangladesh, because landholdings are fragmented. It also represents a financial drain, since all the fuel and hardware must be imported.

Over 50 percent of the land in Bangladesh is irrigated by diesel pumpsets. Almost all of these are owned by the BADC and rented out to groups of farmers at highly subsidized rates. Most of the diesel engines operate low-lift pumps, which lift surface water; the remainder are used in either shallow tubewells (with centrifugal pumps) or deep tubewells (with turbine pumps). However, their operation incurs large costs to the government in the form of subsidies on rental charges and fuel prices, as well as the cost of operating the vast BADC bureaucracy.

Diesel pumpsets are largely underutilized in Bangladesh; on average they irrigate 16 ha when they should be able to cover at least 22 ha of rice and more of wheat. The main reasons for underutilization are the following: the subsidized rental creates little incentive to increase the irrigated area; greater utilization leads to a higher chance of breakdown, which is a risky proposition given the shortage of mechanics and spare parts; also the average landholding is only 0.11 ha, which means that mechanized pumps are being used mainly by large farmers.

Small farmers face several difficulties in organizing themselves into large groups and cooperating for irrigation.

Egypt

Irrigation has been practiced in Egypt for thousands of years using human- and animal-powered lifts. As a result, the technologies used are well adapted to suit the conditions required for pumps in Egypt: lifting large volumes of water (to meet the high crop water requirements) against a very low head. The shaduf is still widely used, but generally only to irrigate small vegetable plots, while the tambour is used to irrigate field crops.

In most areas of Egypt during peak irrigation periods, there appears to be substantial underemployment. Hence the value of small farmers' own labor may be less than the average market wage (i.e., small farmers may not be able to use their own labor in any other way than as a power source on their own land). Small farmers, with less than 2 ha (5 feddans) of land, constitute 94 percent of all farmers in Egypt, and own 57 percent of the cultivated area.

For large farmers, it is expensive to operate a pumping device powered by hired labor, and they are likely to incur a loss on their crops if they use a shaduf for irrigation. For this reason, shadufs are only used by very small farmers who utilize their own labor, usually to irrigate their vegetable plots and other small fields. The shaduf is popular among small farmers also because of its low capital cost (since it can be made from waste material on the farm) and because it involves no risk of failure. The tambour also has a very low capital cost but it is very efficient: with it, one can irrigate eight times as much land as with a shaduf. In many areas of Egypt, small farmers using their own underemployed labor also regard the tambour as a more suitable option than mechanized devices.

In any case, typical small farmers do not have the capital or credit required to invest in mechanized pumping. Mechanized irrigation may force them to cooperate with a large group of farmers, whereas they can use their own tambour and not have to rely on other people to operate or maintain it. Small farmers may also consider mechanized pumping to be risky, since diesel engines often break down and repair facilities are in short supply in Egypt. For such farmers who operate at a subsistence or near-subsistence level, minimizing risk is normally an important consideration.

The main source of power on Egyptian farms (apart from labor) is buffalo and cattle. These animals are also used for breeding and for producing milk, meat, and manure. In 1970, it was estimated that there were over four million cattle and buffalo in Egypt. Almost all the animal-powered pumps used are sakias, though some farmers use tambours fitted with a gear drive and driven by cattle.

It is argued that sakias are too slow to keep us with the demand for water during peak irrigation with the result that crops go short of water. The authors' visit to some areas in Egypt in July 1980, however, revealed that most sakias were underutilized and that their use was not a constraint on production.

Mechanized irrigation is nevertheless being promoted by the government. In 1975, a survey by the Ministry of Agriculture and Irrigation showed that there were 60,000 diesel pumpsets operating in Egypt; most of them were imported and some were assembled in factories in Egypt. Again, the use of diesel pumpsets is economically constrained by the small size of landholdings—the country average is 0.8 ha.[1] Even the smaller 3.8 kW engine can irrigate 14 ha. Some 17 average size farmers would have to cooperate to derive the maximum benefit from such an engine. Large cooperatives are difficult to organize and manage and, for this reason, only large farmers are likely to use diesel pumpsets unless these are heavily subsidized.

A COST COMPARISON OF ALTERNATIVE TECHNOLOGIES

Table 4.1 presents some important characteristics of the alternative technologies being evaluated here. The following points are noteworthy:

- average water output of human-powered lifts is 20–40 m^3/day in Bangladesh and Egypt except for the tambour in Egypt, which gives an estimated output of 173 m^3/day;
- the daily output of animal-powered lifts in Bangladesh is 8–16 times that of human-powered lifts;
- the output of an animal-powered lift (sakia) in Egypt is almost three times the output of another animal-powered lift (chain-and-washer) in Bangladesh;
- daily output of diesel pumpsets (3.8 kW) can be about three times the output of animal-powered lifts, and 35–70 times the output of a human-powered lift.

A petrol pumpset (using a 2.4 kW engine and a 50 × 50 mm pump) is assumed to pump only about 252–360 m^3/day, although this output may be greater with a different pump.

Capital Costs

The capital costs of human-powered lifts are very low ($11–50) except those of the chain-and-washer lift in Bangladesh ($550). The capital costs of animal-powered lifts are about 10–20 percent higher than those of the smallest diesel pumpset (using a 3.8 kW engine). In contrast, a petrol pumpset (2.4 kW) costs about one-third that of a diesel pumpset. About 55–60 percent of the capital costs of animal-powered lifts are the costs of animals (usually cows), which are considerably different at market and shadow prices. About 20 percent of the capital costs of mechanized pumpsets in Egypt and 60–70 percent in Bangladesh are in foreign exchange. These percentages, together with the economic costs of diesel/petrol, affect the economic comparison from the social viewpoint.

Costs of Operation and Maintenance

The operating costs of human-powered lifts comprise mainly labor, and those of animal-powered lifts depend on the costs of feeding animals.

Table 4.1
Some Characteristics of Traditional and Modern Technologies for Low-Lift Pumping in Bangladesh and Egypt

	Bangladesh				Egypt			
	Technology	Water output m³/day	Maximum area irrigable (ha)	Capital costs[1] ($)	Technology	Water output m³/day	Maximum area irrigable (ha)	Capital costs[1] ($)
Human-powered lifts								
- Dhone		19.4	0.2	30	Shaduf	21.6	0.1	11
- Chain/washer		37.0	0.45	550	Tambour	173.0	0.8	50
Animal-powered lifts								
- Chain/washer[2]		324.0	3.1	1,650 (1,000)	Sakia	1,080.0	4.6	1,620 (900)
Diesel pumpset								
- 3.8 kW medium speed		1,378.8	14.0	1,371	3.8 kW medium speed	3,254.0	14.1	1,470
- 6.0 kW medium speed		2,203.0	22.0	1,720	6.0 kW medium speed	5,141.0	22.3	1,846
Petrol pumpset[3]								
- 2.4 kW high speed		252.0	2.5	523	2.4 kW high speed	360.0	1.5	496

[1] At mid-1979 prices converted into U.S. dollars using official exchange rates: $1 = 1 Taka in Bangladesh and $1 = 0.68 LE in Egypt. The foreign exchange component in capital costs is as follows: for Bangladesh, 62–66 percent in the diesel engine and 78 percent in the petrol engine; for Egypt, 19–21 percent in the diesel engine and 20 percent in the petrol engine.

[2] Animals have not been used in Bangladesh, but estimates are presented. Figures in brackets are the cost of cows to operate the lift. Estimated number of cows: two in Egypt and four in Bangladesh.

[3] In Bangladesh, the water output of the petrol engine (2.4 kW) is taken as 7 1/sec as against 38.3 1/sec for the diesel engine (3.8 kW). Similarly, water output from the petrol engine in Egypt is also estimated to be much lower (10 1/sec) than the diesel engine (90.4 1/sec). This is partly because the pump size in the petrol engine is taken as 50 × 50 mm while that in the diesel engine is 100 × 100 mm. Hence, figures on the petrol engines have to be used with caution.

Source: Birch and Rydzewski (1980).

Table 4.2
Annual Operating Costs of Alternative Technologies for Water Pumping ($ per season, mid-1979)

	Bangladesh			Egypt		
			total			total
Human-powered lifts[1]						
Dhone			88.2			-
Chain/washer			88.2			-
Shaduf			-			54.6
Tambour			-			109.6
Animal-powered lifts[2]						
original estimate			23.0 (-12.0)			89.0 (47.3)
revised estimate A			239.4 (204.4)			289.9 (248.2)
revised estimate B			75.9 (40.9)			88.3 (49.6)
	fuel	other[3]	total	fuel	other	total
Mechanised pumpsets						
Diesel (3.8 kW)						
subsidised fuel price	122.1	160.0	282.1	30.9	330.0	360.9
world fuel price	215.1	160.0	375.1	162.4	330.0	492.4
Petrol (2.4 kW)						
subsidised fuel price	202.4	80.0	282.4	91.1	165.0	256.1
world fuel price	252.8	80.0	332.8	190.0	165.0	355.0

[1] Original estimate, assuming hired labor at market wage rate. Interest charges on capital costs and maintenance costs are not included since these would be insignificant, where applicable.

[2] Figures in parentheses are the estimated cost of feed for cows attributed to irrigation. In the "original estimate" this was calculated by subtracting nonirrigation benefits from total feeding costs. In Bangladesh, however, the benefits from milk production and other work ($421) were more than the total cost of feed ($409) ; this explains the minus sign. In "revised estimate A" and "revised estimate B," the feeding cost attributed to irrigation is calculated at 50 percent and 10 percent, respectively, of the total cost of feed.

[3] "Other" costs include lubricating oil, wage of operator, and costs of spare parts and maintenance.

Source: Original estimates are from Birch and Rydzewski (1980, Tables 10 and 11) and revised estimates are based on original data by the editors.

The cost estimates for operating pumping devices are based on one crop in one area of each country. The wage rates used (for human-powered lifts) are local market rates in mid-1979. No allowance is made for interest charges on capital costs for human-powered lifts, since these costs are very low (Table 4.1) and credit is not usually available for such purchases.

In Bangladesh, cattle are used for transport, ploughing and threshing, and as a source of milk and meat. They have not been used for powering irrigation pumps, mainly because of a shortage of fodder and animals. The current cattle population, which survives on grass and available fodder, is found to suffer from a protein deficiency of 30 percent (FAO, 1974). Assuming an average diet consisting of rice straw and *aus* or *aman* crops, a *boro* crop, and some supplemental grazing on nonagricultural land, the cost of the current daily diet can be estimated. The estimated cost of TDN (total daily nutrient) was 0.43 Taka/kg ($0.028/kg in 1979) at market prices. Since a 500 kg cow requires 9.67 kg/day of TDN, its daily feed cost is Taka 4.2 ($0.28). No agricultural land is actually used in the growing of this fodder (since it is available as a by-product), but it is noteworthy that the crop residues from 0.77 ha of agricultural land (double-cropped) are required to feed one cow.

In Egypt, winter feed is based on clover, and summer feeding on wheat straw, maize or sorghum thinnings and leaf strippings, concentrate feeds, and home-grown cereals. Feed supplies are more plentiful in winter, but limited and of poor quality during the summer when there is greater requirement for draught power. Using FAO estimates of nutrient and protein values of food grown in Egypt and on yields of these crops on one typical ha given over to the production of cattle feed for a year (at 19 percent cropping intensity), one finds that 1 ha can produce 4700 kg TDN. Thus, 0.5 ha of land is required to feed one 500 kg cow in Egypt, and its daily diet is valued at $0.68.[2]

Using the above assumptions regarding costs of labor and animals, the costs of operating the traditional lifts in the two countries are estimated. These are displayed in Table 4.2. The original estimate of animal costs was based on the following three-step methodology. First, the cost of feeding the cows was calculated; then, other benefits from cows in terms of value of milk and value of work (other than irrigation) were estimated; last, the cost attributed to irrigation was calculated as the difference between total cost of feed and the sum of estimated non-irrigation benefits. This method resulted in relatively low costs attributable to animals. In the case of Bangladesh, this method implied a negative cost because the estimated non-irrigation benefits from cows were higher than the cost of feed. Since operating costs of animals depend on a number of factors and assumptions, two other cost estimates are presented. "Revised estimate A" is based on the assumption that 50 percent of the total feed costs should be attributed to irrigation, even though cows are used for this purpose for only 63 days a year. This would imply that cows will remain underutilized for about half the time during a season. Furthermore, to account for the fact that dry fodder comes as a by-product of crop production, another estimate is made—"revised estimate B"—assuming only 10 percent of the feed costs attributable to irrigation.[3] Thus the estimated annual operating costs for animal-powered lifts range from $23 to $239 for Bangladesh and from $88 to $290 for Egypt. It may be noted that the costs of animal feed are lower than the costs of labor except in the "revised estimate A".

The operating costs of mechanized systems include: (i) fuel costs, (ii) lubricating oil, (iii) wage of operator, and (iv) costs of spares and maintenance. The fuel costs have been calculated at subsidized market prices (mid-1979) as well as world prices (c.i.f for Bangladesh, which is a net oil importer, and f.o.b for Egypt, which is a net oil exporter).[4] The estimates of

Table 4.3
Total Annual Costs of Pumping Water by Various Technologies (in $)

	Bangladesh	Egypt
Human-powered lifts		
Dhone	92.6	-
Chain/washer	154.2	-
Shaduf	-	56.2
Tambour	-	115.2
Animal-powered lifts		
original estimate	135.5	155.0
revised estimate A	351.9	355.9
revised estimate B	188.4	154.3
Mechanised pumpsets		
Diesel (3.8 kW)		
subsidised fuel price	491.2	471.4
world fuel price	584.0	602.8
Petrol (2.4 kW)		
subsidised fuel price	-	316.5
world fuel price	541.2	415.3

Note: Total annual costs were calculated by adding annual operating costs in Table 4.2 and annualized values of capital costs. The capital costs given in Table 4.1 have been annualized at a 12 percent rate of interest assuming a 12 year life for diesel engine, a 6 year life for petrol engine, and a 10 year life for animals (cows), water wheels, and other equipment. Using official exchange rates: $1 = 1 Taka in Bangladesh and $1 = 0.68 LE in Egypt.

Source: "Original estimates" are from Birch and Rydzewski (1980) and "revised estimates" were elaborated from original data by the editors. See Table 4.2 for explanations.

the prices used are given below, in terms of $/l:

	Bangladesh		Egypt	
	Diesel	Petrol	Diesel	Petrol
Subsidized price	0.176	0.352	0.036	0.128
World price	0.310	0.440	0.189	0.267

Using these prices, the estimated fuel costs for the two alternatives are given in Table 4.2. The diesel oil subsidy was 43 percent in Bangladesh and 81 percent in Egypt. Hence the increase in fuel costs (at economic cost) is significantly higher (about five times) for Egypt. The correction of subsidy in petrol prices raises the fuel cost by 52 percent in Egypt and 25 percent in Bangladesh.

A Comparison of Unit Costs

Using the annualized values of capital costs given in Table 4.1 and operating costs in Table 4.2, total annual costs of pumping water are shown in Table 4.3. The assumptions made in annualization of capital costs[5] are given in Table 4.4, which also shows unit costs in $/1,000 m^3 of water output. The following points may be noted from Table 4.4:

Table 4.4
Summary of Comparison of Costs of Pumping Water by Various Alternatives in $/1000 m^3

	Bangladesh	Egypt
Human-powered lifts		
original estimate (hired labour)	75.70 (Dhone)	33.30 (Shaduf)
	66.10 (Chain/washer)	8.53 (Tambour)
revised estimate (own labour)	3.60 (Dhone)	1.00 (Shaduf)
	28.30 (Chain/washer)	0.45 (Tambour)
Animal-powered lifts		
original estimate	6.63 (Chain/washer)	1.83 (Sakia)
revised estimate A	17.20	4.22
revised estimate B	9.22	1.86

Mechanised pumpsets		
Diesel (3.8 kW)		
subsidised fuel price	5.65	1.85
world fuel price	6.72	2.37
Petrol (2.4 kW)		
subsidised fuel price	25.80	11.27
world fuel price	28.90	14.78

Note: Unit costs in \$/1000 m^3 have been calculated by dividing the total annual costs (Table 4.3) by the volume of water pumped during the season (63 days in Bangladesh and 78 days in Egypt). The volume of water pumped per day for each device is given in Table 4.1. The original estimates in \$/kwh have been recalculated in terms of \$/1000 m^3 by dividing total seasonal costs (capital and operating) by the total capacity or volume of water pumped during the season. Number of days in the season during which water is pumped: Bangladesh 63 days; Egypt 78 days.

Source: See notes and sources of Tables 4.2 and 4.3.

1. Human-powered lifts are the most expensive alternative if the "original estimate" is used, whereby labor is hired at the market wage rate of \$0.70/day in each country. However, this wage will vary from season to season and across regions. Moreover, if small farmers can use their own labor or that of a family member, the value of this labor is not necessarily equal to the market wage. In areas of high unemployment, as in Bangladesh throughout the year and in Egypt during the peak irrigation seasons, such farmers may not find alternative employment, so that their labor has little or no value in market terms. Thus, from the farmers' standpoint, the economics of using human-powered lifts critically depends on the farm size and level of unemployment or underemployment in the region or season.

2. Under the assumption that a small farmer or a family member will operate a human-powered lift for about six hours per day without affecting output elsewhere, the real resource costs of labor can be taken as zero. This assumption leads to the unit costs given under the "revised estimate (own labor)" in Table 4.4. One might note that the dhone in Bangladesh and both the shaduf and tambour in Egypt turn out to be the least-cost alternatives for low-lift irrigation.

3. Even when a large farmer is operating with hired labor, the market wage does not represent the economic cost of using this labor from the society's standpoint. Under conditions of severe unemployment and/or underemployment, the social opportunity cost of unskilled labor (i.e., the social benefit forgone from an alternative use for labor time) may be very low or even zero. Thus, in terms of real resource cost to society, using labor for low-lift irrigation may be more economic than the alternatives considered here.

4. Animal power is more expensive than the diesel engine in Bangladesh, even under the assumption that only 10 percent of the feed costs are attributable to irrigation. Animal-operated lifts are economic when fodder costs are taken as zero and diesel oil is priced at world prices. In Egypt, however, animal-powered lifts ("original estimate" and "revised estimate B") cost about the same as diesel pumpsets when diesel oil is priced at its subsidized rate. When the world price of diesel is used, animal-powered lifts are economic under the assumption that only 10 percent or less of the total feed costs of cows are attributed to irrigation. The latter assumption needs to be looked at carefully. Thus, the economics of animal power are site-specific, and it may be economically and financially profitable, particularly in areas where cattle can be fed on crop residues, pastures, and forest land.

5. It is noteworthy that the costs of cows are about 50–60 percent of the total capital costs of animal-powered devices. If it is assumed that fodder has zero opportunity cost in some regions, then the real resource costs of raising cows there may also be zero or very low. Under these assumptions, the use of animals for lift irrigation will be generally economical from the viewpoint of society.

6. The unit cost of water pumped by diesel pumpsets increases by 20 percent in Bangladesh and 28 percent in Egypt when world prices are used instead of subsidized prices. This adjustment makes it less profitable to use a diesel pumpset compared with the sakia or the animal-driven chain-and-washer lift.

7. The command area of a diesel pumpset is 14 ha, which is rather large considering the size distribution of farm holdings in Bangladesh and Egypt. In Bangladesh, approximately 70 percent of the farmers own less than 1 ha; together, they cultivate about 28 percent of the total cultivated area. Furthermore, about 20 percent of all farmers own between 1 and 3 ha and cultivate about 40 percent of the total cultivated area in the country.[6] Similarly, in Egypt, small farmers who own less than 2 ha (five feddans) of land constitute 94 percent of all farmers and own 50 percent of the cultivated area. Also, about 10 percent of the Egyptian farmers own between 2 and 4 ha of land. These data on the pattern of landownership may provide a useful guide for the choice of low-lift irrigation devices.

8. Given the land distribution, only a few large farmers will find it economic to use a diesel pumpset. For small farmers, this option will be underutilized for a major part of the year. For example, for Bangladesh, if a comparison is made between an animal-powered lift and a diesel pumpset for a 3 ha farm, the former will be fully utilized but the diesel option will be only partially utilized. The quantity of water that is pumped by the animal-powered lift in 63 days will be pumped by the diesel pumpset in less than 15 days. Hence, the diesel pumpset is likely to remain idle for three quarters of the time unless the water is sold to other farmers or the pumpset is

cooperatively owned by four or five farmers each owning around 3 ha. If these conditions are not met, the diesel pumpset will remain idle.

9. The petrol pumpset is not found to be economic because of the low discharge of water (due to small pump size), lower efficiency of the petrol engine (petrol consumption of 0.38 1/kWh as against diesel consumption of 0.29 1/kWh) and higher fuel prices (at both market and shadow prices).

A Comparison of Present Value of Costs for a Given Farm Size

Annual unit costs estimated in $/1,000m^3 can be used for comparing alternative technologies if it is assumed that each of these technologies is fully utilized. The diesel pumpset can be fully utilized on a 14 ha farm where water is pumped every day for 10 hours for a period of 63 days in a season

Table 4.5
A Comparison of Present Values of Costs of Irrigation Using Alternative Technologies ($)

	Bangladesh			Egypt		
	Small up to 0.5 ha	Medium 3.1 to 4.6 ha	Large 14 ha	Small 0.8 ha	Medium 4.6 ha	Large 14.1 ha
Human-powered lifts	75 (own labour; lift)	3,281 (hired labour; 6 lifts)	15,311 (hired labour; 28 lifts)	50 (own labour; 1 lift)	5,668 (hired labour; 6 lifts)	16,060 (hired labour; 17 lifts)
Animal-powered lifts (original estimate)	1,793 (1 lift)	1,793 (1 lift)	7,172 (4 lifts)	2,172	2,172	6 516
Diesel pumpset						
Subsidised fuel price	2,390	2,530	3,120	3,457	3,578	3,707
World price	(2,410)	(2,659)	(3,696)	(3,587)	(3,951)	(4,520)

Note: The interest rate used is 12 percent. Assumptions about life of equipment were given in Table 4.3. One human-powered lift is assumed to irrigate 0.5 ha in Bangladesh and 0.8 ha in Egypt. It is assumed that a diesel pumpset (with a 3.8 kW engine) is underutilized on small farms and fuel costs are proportionately lower.

Source: Elaborated from data in Tables 4.1 and 4.2.

in Bangladesh. This yields a total water output of 205,000 m^3 per season or 14,643 m^3/ha per season. However, if the same diesel pumpset is considered for a smaller farm size (e.g., 3.1 ha), its utilization rate drops because there is not sufficient demand for water. Assuming that the possibility of selling water is limited, this means that only about 45,390 m^3 of water may be pumped during the season and the utilization of the diesel pumpset is only 22 percent. In other words, instead of pumping for 630 hours (63 days for 10 hrs/day), it pumps for 140 hours (the number of days depends on water requirements). Of course, the fuel costs are proportionately lower. But since fixed costs (including capital costs) are not lower, the present value of total costs does not significantly decrease. Similarly, the utilization of diesel pumpsets is even lower on a small farm (0.5 ha) and this means high fixed costs per ha. One conclusion that must be drawn from this discussion is that a true comparison of costs requires that each technology is assessed for a given farm size and not under the assumption of varying farm sizes.

Table 4.5 presents the results of this type of analysis for small, medium, and large farmers in the two countries. It is further assumed that small farmers use their own labor (and family labor) for operating lift-irrigation systems, while medium and large farmers depend on hired labor. The comparison is also in terms of the sum of present values of capital and operating costs of each technology.

SUMMARY OF CONCLUSIONS

As shown, the human-powered lifts are the most attractive technology for small farms where the farmer's own labor is used. For medium farms (3.1–4.6 ha), animal-operated lifts are economic, even when the full capital costs of purchasing the cows are taken into account. Due to capacity underutilization, the present costs of a diesel pumpset (3.8 kW) are higher than those of animal-powered lifts for medium farms in the two countries. For large farms, however, the present costs of the diesel pumpset are the lowest because capacity utilization is higher. For these farms, the costs of human- and animal-powered lifts are higher because hired labor and animal feeding are too expensive.

In the final analysis, the economics of traditional sources depend crucially on the farm size in question and the real resource costs of maintaining animals on the farm and the pattern of their utilization. Diesel pumpsets are economical only for large farms that do not maintain animals for non-irrigation purposes.

NOTES

1. See Government of Bangladesh (1978) for details on landownership and farm size.

2. Data from FAO (1974 and 1977); see Birch and Rydzewski (1980) for details.

3. An even lower percentage of the cost of feed requirements may be reasonably attributed in situations where cattle, for example, are used primarily for other purposes such as milk or meat production and are underutilized. In such situations, irrigation may not require additional animals or reduce the non-irrigation output of existing animals.

4. World fuel prices are used here as an approximation of the economic cost of fuel. However, if a social cost/benefit analysis were carried out for the different technology options, the price of imported fuels should be converted into local currency using the shadow exchange rate. Wages should be shadow priced to reflect social opportunity costs, and import duties or subsidies on capital goods should be removed. However, the data in the original study do not permit this type of analysis.

5. No data were available on the life of centrifugal pumps, but the pump body should have an almost indefinite life. The authors assumed a life of 20 years for these pumps in Bangladesh and Egypt (allowing for maintenance of bearings, glands, and possible replacement of a damaged rotor). The FAO/World Bank (1978) assumed that each pump owned by BADC operated 384 hours per year, thus giving the pump a life of 17–18 years (assuming an average life of 6,500–7,000 hours). However, a pump used to its fullest to irrigate *boro* rice in Comilla would operate for 630 hours per year. The reduced life expectancy (assumed by the authors) for the diesel engine is 12 years for both Bangladesh and Egypt. Although the Egyptian engine operates for a higher number of hours, the same life is estimated because renting of pumpsets (as opposed to owning them) is more common in Bangladesh and implies a higher rate of damage. Similarly, the life expectancy of a petrol engine was estimated at six years in Egypt. The interest rate used in annualization was 12 percent.

6. The irrigation decision of a farmer choosing between a human-powered device and a diesel pumpset available for hire by a cooperative has been discussed as a management training case in *Rural Technology Choice.* (see Reynolds, 1981).

5

Windmills for Irrigation: Sri Lanka, Kenya, Cape Verde, and the Sudan

Consultancy Services Wind Energy Developing Countries (CWD), summarized by Ramesh Bhatia and Madhura Swaminathan

Wind energy has a long history of utilization, from the use of sail to propel boats several thousands of years ago to the use of windmills to grind maize, lift water, and generate electricity. The increase in the cost of conventional energy resources over the 1973–1983 decade led to a revival of interest in wind energy. Many developing countries have begun to assess potential for the application of windmills for development needs. This chapter is a summary of four case studies on the feasibility of using wind energy for water pumping in Sri Lanka, Kenya, Cape Verde, and the Sudan.

The potential for wind energy development depends primarily on wind frequency and wind speed, energy output being proportional to the cube of wind speed. Variability in wind speed, whether during the day, month, or year, affects the quantity of water pumped which, in turn, affects cropping patterns, need for storage tanks, and so on. The economics of a windmill for irrigation pumping depends crucially on the wind regime, which determines the size of the windmill, design features, seasonality of water output,[1]

This chapter is a summary of the following case studies: J.A.C. Vel and L.R.v. Veldhuizen: *A model for the economics of small scale irrigation with windmills in Sri Lanka* (Amersfoort, The Netherlands, CWD, mimeographed series, Oct. 1981); W.E. van Lierop and L.R.v. Veldhuizen: *Wind energy development in Kenya* (ibid., Nov. 1982); H.J.M. Beurskens: *Wind energy for water pumping in Cape Verde* (ibid., Feb. 1981), and Y.H. Hamid and W.A.M. Jansen: *Wind energy in the Sudan* (ibid., July 1980).

and so on. This makes windmills rather location-specific and case studies are required to find suitable sites for their use.

All four country studies discussed here compare the use of a windmill for water pumping with at least one other alternative. A windmill is compared with a kerosene pumpset in Sri Lanka, with a diesel pumpset in Kenya and the Sudan, and with diesel, solar-photovoltaic, and solar-thermal systems in Cape Verde.

The different technologies are compared in terms of their total unit costs per m^3 of water output. This method is based on the assumptions that (i) the windmills and the competing technologies are equivalent in the sense that each provides the same quantity and quality (time pattern) of water, and (ii) their benefits are identical.

The studies differ in methodology of estimating costs and in their comparison. For Sri Lanka, costs are estimated both at financial prices (or market prices) and economic prices (excluding subsidies) to reflect differences between the viewpoint of users and that of society. The studies of Sri Lanka and Cape Verde also consider the foreign exchange expenditures on alternative technologies to reflect the impact on the balance of payments. Employment estimates for local production of windmills are given in the study of Cape Verde. Costs per m^3 of water output were converted into U.S. dollars by using the official exchange rates prevalent at the time (early 1980 and mid-1981 for Kenya). For Sri Lanka, total annual costs, which include annual installments on capital cost, repair and maintenance costs, and fuel costs (for kerosene pumpset) were also compared. None of the studies used life-cycle costs or the present value of the sum of capital and operating costs of alternative systems.

In the Sri Lanka study, apart from cost comparisons, an attempt was also made to estimate the net benefits of using irrigation water. Annual farmers' income was calculated on the basis of the value of agricultural output and farm costs. These data were used to calculate the economic internal rate of return (EIRR) for pumping systems using windmills and kerosene based on economic prices. The Kenya study also attempts to estimate the benefits of lift irrigation under varying conditions of crop mix, farm size, and water availability.

It would be tempting to try a comparison of costs per m^3 of water output by windmills in different countries. However, such a comparison would not be realistic or useful for the following reasons: (i) windmills suitable for one location may not be suitable for another; (ii) wind regimes as well as capital costs, interest rates, and fuel prices vary across countries and regions; and (iii) approaches used to study the potential of windmills, land-use patterns, and other technical and agronomic assumptions differ from one study to another. Hence, this chapter is divided into four sections that review the individual country studies.

SRI LANKA

Taking into account technical, agricultural, and socioeconomic aspects, the Wind Energy Utilization Project of the government of Sri Lanka has developed a windmill—the small WEU I/2—that is suitable for irrigation in the Dry Zone. It has a rotor diameter of 3 m and a total head of 7 m.

For the average farmer with about 0.8 ha of land, the WEU I/2 windmill allows the farmer to cultivate 0.4 ha in the *Yala* season (April to August). Since wind is a relatively uncertain factor, irrigation water may not be available everyday; thus a storage tank has been added to the windmill. An alternative device for irrigation is a 2-inch centrifugal pump powered by a kerosene engine with a capacity of 2 hp and a maximum daily output of 55 m^3, pumping 6 hours a day and using 1.13 1/hr of kerosene. The crucial difference between a windmill and the kerosene pumpset is that the latter can irrigate a larger command area if groundwater is available.

Several alternative cropping patterns are open to the farmer (e.g., planting crops more sensitive to drought as water supply is more guaranteed), but the one considered here is where the cropping pattern is identical for the windmill and the kerosene pumpset. Several criteria can be used to indicate the financial attractiveness of a windmill for irrigation, including net income benefits from cultivation in the dry (Yala) season, management skills and risk, and total annual pumping cost per m^3 of water used for irrigation. The last criterion may be the most attractive one to use in financial and economic comparisons of alternative technologies for irrigation.

From the viewpoint of society, two additional criteria can also be used: the EIRR, which indicates the profitability of investing capital in different alternatives, and the costs in foreign exchange, which is normally scarce. The important differences between financial and economic analyses are:

1. difference between market prices and real prices of equipment and inputs, because of subsidies (e.g., on fertilizer, 75 percent; kerosene, 38 percent), market imperfections such as monopolies, and inefficiencies in distribution; and and
2. the real resource cost of unskilled labor may be lower than the market wage rate due to unemployment/underemployment.

In the analysis reported here, only one correction has been made to the original study: in the financial analysis, the price of kerosene has been taken at Rs 3.42/1, while in the economic analysis it is taken as Rs 5.5/1. Estimates of windmill output are based on wind data collected in 1980 by the Wind Energy Unit in Morazewa (multiplied by a factor of 1.2 for the Dry Zone, which is considered to be representative). Availability of groundwater is another important factor in determining choice of windmill and its output. There is a danger that wells will become dry at the end of the Yala season (August) in the Dry Zone, which is disadvantageous to the windmill farmer since July and August are the most windy months. Here it is assumed that water is available throughout the Yala season. The farmer buying a windmill must build a storage tank to collect small amounts of water when there is limited wind, and to save the water pumped by the windmill when he is not able to use it, especially at night. A cheap tank with a capacity of 22.5 m^3 (50 percent of maximum average daily output) is considered here. In addition, earth channels will be required to operate the surface irrigation system (nowadays plastic tubes are also used). Further, there is a possibility of using the windmill in the *Maha* season, which is not considered in this study due to lack of data.

All prices used are based on information on actual market prices for 1980. Though not realistic, inflation is assumed to be at the same rate in all sectors of the economy. The only relative price increase that is taken into account is that of kerosene fuel: 5 percent per year, which is lower than the actual increases over the 1981–1985 period. Since a windmill farmer requires a larger initial investment as compared with a kerosene pumpset, the farmer selecting a windmill can benefit from inflation. The credit facilities available make it very difficult for small farmers who lack securities to borrow for agricultural investments. Here it is assumed that banks will adapt their rules and give loans for windmills, which are to be repaid at a compound interest rate of 14 percent.

Comparison of Windmill and Kerosene Pumpset

The two alternatives have been defined in such a way that the level of agricultural benefits is identical in the two situations. An area of 0.4 ha is cultivated in the Yala season with chillies, onions, pulses, and vegetables. The total gross irrigation requirement (i.e., net irrigation requirement divided by irrigation efficiency) is 5050 m^3 per year and, given the windmill output, the possible command area for cultivation is estimated. Using market prices, the gross production value of this cropping pattern is estimated at Rs 12,225, or $764.

Tables 5.1 and 5.2 summarize the major differences between the two alternatives. The capital costs of the windmill (excluding well and field channels) are almost three times those for the kerosene pumpset. However their annual costs are almost equal, because of the significant cost of kerosene. Using the annual cost to farmers of pumping per m^3 water as the criterion for comparison, one finds that the kerosene pumpset performs only slightly better than the windmill. But the economic comparison turns out in favor of the windmill, mainly because of the subsidy on kerosene, which was about 38 percent. Furthermore, the windmill requires less foreign currency and possibly generates more employment.[2]

Table 5.1
Costs of Windmill and Kerosene Pumpset for Irrigation in Sri Lanka ($, 1980 prices)

	Windmill	Kerosene pumpset
Capital costs		
windmill	750.00	211.86
installation	181.25	85.00
tank costs	156.25	-
Subtotal capital costs	1,087.50	296.86
Total capital costs including well and channels[1]	1,462.50	671.86
Annualised capital costs[2]	280.74	129.00
Annual repair and maintenance costs	62.81	28.12
Fuel costs[3]	-	-
financial analysis	-	168.88
economic analysis	-	273.93
Total annual costs[4]		
financial analysis	343.55	326.00
economic analysis	343.55	431.05

Note: Exchange rate used: Rs 16 = $1.
[1] The costs of well and field channels were Rs 3000 ($187.50) each.
[2] Annual installment = total investment cost × annuity factor where annuity factor is 0.192 (i = 14, n = 10) given by $i(1 + i)^n/(1 + i)^n - 1$.
[3] The financial price of kerosene = Rs 3.42/l; and its economic price = Rs 5.5/l.
[4] Total annual costs = annual installment on capital cost plus repair and maintenance cost plus fuel costs.

Source: Elaborated from Vel and van Veldhuizen (1981, pp. 56–73).

A sensitivity analysis is carried out with respect to the key parameters (i.e., interest rate, relative price increase of kerosene, and subsidies on kerosene and fertilizer). Since windmills have a high initial investment, the results are very sensitive to interest rates. The windmill is found to be the most attractive option on the basis of the criterion of pumping cost per m^3 when the interest rate is below 10 percent. Furthermore, with a price increase for kerosene of more than 8 percent, the windmill is financially attractive to a farmer. The effect of government subsidies (e.g., on kerosene) is an important determinant of costs to a farmer. At economic or nonsubsidized price of kerosene, irrigation water from the kerosene pumpset costs nearly 27 percent more than that from a windmill (Table 5.2).

Table 5.2
Comparison of Windmill and Kerosene Pumpset for Irrigation in Sri Lanka ($, 1980 prices)

	Windmill ($)	Kerosene pumpset ($)
Annual farmer's income	465.31	482.75
Extra income per man-day in Yala	1.47	1.55
Pumping cost per m^3 of water[1]		
financial analysis	0.068	0.064
economic analysis	0.068	0.085
Economic internal rate of return = (EIRR)[2]	35.7%	45%
Annual foreign currency expenditure[3]	161.62	291.75

[1]It is assumed that both options would pump 5050 m^3 of water per year and that 50 percent of the farm area is used in the Yala season. The exchange rate was $1 = Rs 16; the financial price of kerosene = Rs 3.42/l and the economic price = Rs 5.5/l.
[2]The EIRR is the interest rate at which the present worth factor for uniform series is equal to the ratio of capital investment and the difference between gross production value and annual recurrent cost.
[3]For the windmill, it is assumed that 75 percent of materials are imported, 1.5 percent and 35 percent of installation and maintenance costs, respectively, are in foreign currency, and 100 percent of fertilizers used are imported. The kerosene pumpset is assumed to require 65 percent foreign currency for initial investment, 35 percent for maintenance and 90 percent for fuel consumption. Agricultural inputs are the same as for the windmill.
Source: Elaborated from Vel and van Veldhuizen (1981, pp. 56–73).

KENYA

Although Kenya has no well-established tradition in small-scale irrigation, the need to promote agricultural growth has led to a renewed interest in very small-scale lift irrigation. The case study summarized in this section seeks to explore whether windmills will be appropriate for small-scale irrigation projects. Before discussing windmills, we first need to take account of the general socioeconomic considerations that make irrigation feasible. In general, lift irrigation, with water supplied by some kind of pump, is only marginally viable in most parts of Kenya. If water is more than 10–15 m below ground surface, the cost of each m^3 of water pumped is too high to be used in a profitable way in irrigation (given present cropping patterns, prices, etc.). The only way to increase the profitability of lift irrigation is the introduction of high-value crops (such as vegetables) in the cropping pattern, but this requires infrastructure for input supply and marketing.

Using information on wind conditions and irrigation requirements, the command area for cultivation can be estimated. These estimates vary for different regions, and only two regions are considered here. In Kisumu, the area it is possible to cultivate with a 12PU500 windmill is 0.6 ha for a head of 10 m (apart from three months of the rainy season). In the Lower Tana region, where annual average wind speed is higher and the wind distribution throughout the year is more uneven, it is possible to irrigate a relatively larger area (3.2 ha) from April to October and a very small area (0.7 ha) from November to March.

The economic evaluation of windmills depends largely on assumptions regarding the following: wind speeds and their conversion to useful water output, capital costs, inflation rate, interest rate, and price of fuel. The two major methods for economic evaluation of projects use either the total present value of costs or annual costs over the productive lifetime of investments. The latter is used here to compare windmills with diesel pumpsets in Kisumu and Lower Tana.

Kisumu Region

A study of windmill irrigation in the Kano plains, south of the city of Kisumu and along Lake Victoria, was carried out. Assuming vegetable cultivation throughout the year, a suitable windmill (a 12PU500 unit) would have a command area of 0.6 ha from April to October and 0.5 ha in other months, at a total head of 10 m (see Table 5.3). The average annual costs of this system were estimated and compared with those of a possible alternative: a diesel pumpset. At 1981 prices, the total investment costs were K.Sh 35,690 ($3399) for the 12PU500 windmill including storage tank, and K.Sh 41,300 ($3933) for the diesel pumpset (a 5.25 hp diesel engine with centrifugal pump). For the diesel pumpset, two cases are considered: under a high utilization rate, the pumpset is assumed to work 1000 hr/yr with a total output of 47,000 m^3 and a command area of 2.7 ha; under a low utilization rate, water output is 13,000 m^3. Table 5.4 shows that under the low utilization case, the diesel pumpset is less attractive than the windmill in terms of costs per m^3.

However, a true cost picture can only be obtained if the sensitivity of the results to variations in underlying assumptions is attempted. For instance, a fuel price escalation rate of 10 percent will raise the costs of using a diesel pumpset to K.Sh 0.77 ($0.073) and K.Sh 0.45 ($0.042) for the low and high utilization cases, respectively. At higher interest rates, the windmill will compare more favorably with the diesel pumpset as the latter requires

Table 5.3
Some Characteristics of Windmills in Kisumu and Lower Tana Regions, Kenya

	Kisumu	Lower Tana
Wind potential	Low	Medium
Windmill		
type	12PU500	12PU500
diameter	5 m	5 m
pumping head	10 m	2.8-5.3 m
Cropping pattern	Vegetables	Rice and maize
Windmill output	24 312 m^3/year	75 720 m^3/year
Minimum command area	0.5 ha	0.7 ha
Water requirements	13 000 m^3/year	56 000 m^3/year
Storage capacity	21 m^3/day	21 m^3/day
Lifetime	10 years	10 years

Source: Lierop et al. (1982, pp. 102–110).

higher investments. If wind speeds are higher than those assumed here (as is likely in a few places along Lake Victoria), then a resulting increase in windmill output will reduce the cost per m^3 of water pumped.

Lower Tana Region

Along the lower and middle reaches of the Tana River, there is a certain tradition of irrigated agriculture, for which a 12PU500 can be suitable. Assuming a cropping pattern of rice, maize, and some other highland crops, the command area is about 3.2 ha in the season of the long rains and 0.7 ha in the other season. Total annual water requirement is 56,000 m^3, which is 74 percent of predicted windmill output. The costs of water pumping calculated by the method of annual values (Table 5.4) suggest that the windmill is more attractive than the diesel pumpset. All costs per m^3 are much lower in this region than in Kisumu because of much lower head requirements. The reduction in investment costs per m^3 of the windmill is more than proportional to the decrease in head because wind conditions are also more favorable. Total costs per m^3 are also lower for the windmill even when the diesel pumpset is considered under the high utilization case (i.e., where 110,000 m^3 of water is pumped, enough to irrigate 3.5 ha throughout the year).

Table 5.4
Comparison of Costs of Water Pumping by Windmill and Diesel Pumpset for Irrigation in Kisumu and Lower Tana ($/m^3 at 1981 prices)

	Kisumu area		Lower Tana	
Windmill				
Instalment on capital				
main investments (n=10, i=7%)	0.026		0.006	
storage tank (n=10, i=7%)	0.010		0.002	
Operation, maintenance, repair and lubrication	0.014		0.003	
Total costs/m^3 of water	0.050		0.011	
Diesel pumpset[1]	Utilisation		Utilisation	
	Low	High	Low	High
Instalment on capital costs (n=10, i=7%)	0.042	0.012	0.010	0.005
Average fuel costs[2]	0.019	0.019	0.008	0.008
Operation, maintenance, repair and lubrication	0.004	0.004	0.002	0.002
Total costs/m^3 water	0.066	0.036	0.020	0.015

Note: Total water utilized is 13,000 m^3/year in Kisumu area and 56,000 m^3/year in Lower Tana. The low utilization case of the diesel pumpset (see note 1) corresponds to the amount of water pumped by the windmill. The exchange rate used is $1 = K.Sh 10.50.

[1] It is assumed that the crop pattern is the same for the windmill and the diesel pumpset (with a 5.25 hp engine). In Kisumu, the diesel pumpset supplies 13,000 m^3 of water in the low utilization case and 47,000 m^3 in the high utilization case. In Lower Tana, it supplies 56,000 m^3 in the low utilization case and 110,000 m^3 in the high utilization case.

[2] Assuming a fuel price increase of 5 percent a year.

Source: Elaborated from Lierop et al. (1982, pp. 102–110).

Also in terms of benefits, the performance of lift irrigation seems to be better in Lower Tana than in Kisumu due to availability of water at shallow depth. The variability of windmill output has no serious effect as no high-value vegetables (which are sensitive to water shortage) are included in the crop plan. The most serious drawback of the windmill, however, is the smaller cultivated area (as compared with a farm using diesel pumps), especially in the off-season. This effect becomes less pronounced in areas closer to the coast.

An analysis incorporating economic considerations also needs to be done. Though more difficult to quantify, the use of shadow prices for traded goods or of a standard shadow exchange rate should be tried. For instance, windmills are locally produced but use imported materials, whereas the total requirement of diesel for pumps is imported. If the shadow price of foreign currency is assumed to be about 40 percent higher than the official exchange rate, the windmill compares more favorably. Also windmill production in a factory at Thika (Kijito) employs more than 35 persons on a full-time basis. An increased utilization of these locally made windmills would result in additional employment in Kenya.

Thus windmills produced in Kenya are attractive for small-scale irrigation at relatively low volumes of water pumped annually. Local studies have also shown that, at moderate wind regimes, windmills can pump water for drinking-water supply at nearly half the cost of diesel pumpsets. In the Kisumu region, the 12PU500 windmills are able to pump at the lowest cost per m^3 of water (i.e., at their optimal utilization) in cultivated farms of about 1 ha and total lift of 10 m. For farms above 2 ha, diesel pumpsets operate at lower costs.

On the basis of cost comparisons, the 12PU500 windmill is an attractive water-lifting device along the Lower Tana region. However, lower wind speeds in the season of short rains seem to be a drawback for the use of windmills in this area. The analysis with shadow prices suggests that local windmills compare even more favorably with diesel pumpsets from the society's viewpoint—that is, the cost difference between the windmill and the diesel pumpset increases if, instead of market prices, shadow prices are used.

CAPE VERDE

The case study summarized in this section concerns the feasibility of using windmills in rural areas for irrigation, cattle watering, and domestic water supply. The only source of water for irrigation in Cape Verde is groundwater. During most of the year, the islands are affected by the trade winds from the northeast. Three wind seasons can be distinguished: January–June, July–September, and October–December, with average wind speeds of 7.7, 4.7, and 6.6 m/sec, respectively. There are two kinds of windmill sites in the islands: the *ribeiras* (valleys, lowlands) where agriculture is mostly undertaken, and the *achadas* (highlands, plateaus). The average wind speed in the ribeiras, facing northeast, is about 70 percent of that in the achadas.

The study carried out a comparison of financial costs of a windmill with those for a diesel pumpset, a solar-photovoltaic system and a solar-collector rankine system. In the comparison of costs, only total system costs are taken into account, excluding operators and watchmen. A storage tank is necessary for windmills because wind energy supply occurs irregularly. The tank capacity is assumed to be equal to the water consumption of two days, which accords with local experience. The cost comparisons are based on a common annuity method, which includes annual capital costs, maintenance costs, fuel costs, and operating labor.

Two typical situations are selected for the windmill: one where a windmill directly driving a piston pump is installed on a well in a ribeira, with an average total pumping head of 20 m and average windmill diameter of 3.5 m; another where the same system is on a tubewell on an achada with an average total pumping head of 76 m and an average windmill diameter of 3.7 m. Both cases use the CWD prototype windmill model THE-II.

Some characteristics and costs of windmills are presented in Table 5.5 while Table 5.6 compares the unit costs of pumping water with the alternative technologies considered. For windmills, one finds that the cost of tanks

Table 5.5
Technical Features and Costs of CWD Windmill[1] in Cape Verde ($, at 1980 prices)[2]

	Ribeiras	Achadas
Wind speed (annual average)	4.7 m/sec	6.7 m/sec
Total pumping head	20 m	70 m
Annual water yield	16,000 m^3	13,700 m^3
Capital cost	1,259.0	1,396.7
Installation	625.0	625.0
Miscellaneous	81.9	81.9
Total capital costs	1,965.9 (247.7)[3]	2,103.6 (265.8)[3]
Operating and maintenance costs	205.5	212.5
Total costs per m^3 of which:	0.080	0.087
windmill cost	0.028	0.035
tank cost	0.052	0.052

[1]The CWD windmills (THE-II) are locally produced and have the following diameters: 3.5 m in ribeiras and 3.7 m in achadas.

[2]Currency conversion at official rate: $1 = 40 escudos.

[3]Figures in parentheses are for annualized values assuming i = 10 percent and a life of 15 years.

Source: Elaborated from Beurskens (1981).

Table 5.6
Comparison of Costs of Pumping Water by Alternative Technologies in Cape Verde ($/m³ at 1980 prices)

Technology options	Ribeiras	Achadas
Windmill (a) locally made	0.080	0.087
(b) imported	0.086	0.094
Diesel pumpset[1] (5.2 kW engine[2] at $0.375/l for fuel)	0.082	0.153
Solar photovoltaic system[1] (peak power = 220 W)[3]	0.49	1.73
Solar collector rankine system[4] (using Sofretes solar pump) (peak power = 325 W; efficiency = 0.57)	0.90	3.28

Note: At official exchange rate: $1 = 40 escudos.

$$\text{Annual capital costs} = \frac{PV(1 + i)^n}{(1 + i)^n - 1}$$

where i = interest rate, n = working life (years); and PV = present value of initial investments.

[1]Energy needed to pump the same amount of water as windmills do in the two situations.

[2]The diesel pumpset costs $3812.5 (including installation) and is assumed to have a life of 10 years (low utilization). Interest rate = 10 percent. A small diesel pumpset produces 1 kWh water energy, consuming 1.0 liter of diesel oil.

[3]The cost of photovoltaic system is taken at $6000, based on information from Solar Electric International. If 100 machines are purchased, installation costs equal $625. A life of 15 years is assumed. The efficiency defined as water power = peak power = 37 percent.

[4]Based on the Sofretes pump, which costs $47,500 (including installation) and has a service life of 15 years.

Source: Elaborated from Beurskens (1981).

is more than 50 percent of the unit cost of water pumped. It is therefore advisable to investigate possibilities for reducing the capital costs of tanks. The tables suggest that windmills, especially those made locally, are competitive with diesel pumpsets. The costs of both diesel pumpsets and wind pumps are much lower than the costs of solar systems.

From a national perspective, savings in foreign exchange and job creation are important. The foreign exchange component includes both costs of imported capital goods and fuel. Beurskens (1981) estimated that locally produced windmills (with a 5 m diameter) would cost over $2000 less than imported models and, if the diameter can be reduced to 4 m, the currency savings would be about $1300 per windmill. The installation of a windmill for irrigation would save about $325 per year in diesel oil at c.i.f prices of 1980. Local production of windmills, instead of imports of windmills or diesel pumpsets, creates additional employment. Labor needed for installation and maintenance will be roughly the same for both locally produced and imported windmills. The extra employment created in the production of a 5 m CWD prototype is about 550 worker-hours, including 250 worker-hours of low skilled labor, 250 worker-hours of highly qualified labor and about 50 worker-hours of management personnel. For small-scale production of 4 m windmills, the labor required is around 450 worker-hours.

THE SUDAN

An ex-ante study of the potential for windmills in the Sudan is summarized here. Geographically, the major part of the Sudan is a huge plain crossed by the Nile and its tributaries; it has a desert in the north and semi-desert in the northeast. North of about latitude 19°N, northerly winds prevail for most of the year and rainfall is rare; to the south, the seasons are characterized by the oscillation between moist southerly air and dry northerly air. Average wind speeds around 4 m/sec, stable winds, large open areas, and low population densities suggest a potential for wind energy utilization.

In the Sudan, one has to consider the occurrence of gusts and *haboobs* (strong whirlwinds) when designing the windmill structure. Thunderstorms may be accompanied by severe ground squalls in which wind speed, particularly in the north and central Sudan, may exceed 110 km/hr on occasion. The increase of wind speed and change in wind direction are both sudden and the windmill should be strong enough to withstand this.

Further calculations and cost comparisons are based on data for Khartoum, which is considered to be representative of large parts of the Sudan. The average wind speed is 4.3 m/sec; there is very little rainfall (180 mm) and, without irrigation, only one crop can be cultivated in a year. Water requirements for irrigation are relatively large. There are two possibilities for water irrigation in the area of Khartoum (Nile Valley): using ground water from tubewells with water levels around 20 m underground, and using surface water to be pumped from the Nile. The most suitable windmill for these requirements is a 7 m diameter machine and the feasibility study is based on this model, assuming that the 5 m CWD prototype can be scaled up. Windmills are compared with diesel pumpsets, which are the predominant technology for pumping in these areas.

The financial cost analysis is made for the hypothetical 7 m diameter windmill (Khartoum wind data). The rate of interest used is 9 percent, which is the rate at which the Agricultural Bank provides medium-term loans. The inflation rate considered is 6 percent, which means that the costs of operation, maintenance, and fuel are expected to rise by at least that rate. Windmills are designed for a lifetime of 10 years, while the expected lifetime of the diesel pumpset is 7 years.

Table 5.7 presents the costs per m^3 of water for irrigation to farmers using a windmill and a diesel pumpset for an output of 73,000 m^3 in both cases.

Table 5.7
Costs of Water Pumping for Irrigation with Windmill and Diesel Pumpset in the Sudan ($, at 1980 prices)

Cost items	Windmill[1]	Diesel pumpset[2]
Capital cost	3,625	3,250
Installation	125	-
Total capital costs	3,750	3,250
Storage tank (capacity = 260 m^3)	3,750	-
Operation, maintenance and repair	187.5	375.0
Annual water output	73,000 m^3/yr	73,000 m^3/yr
Annual fuel costs at £0.167/l or $0.33/l	none	422.55
Financial costs/m^3 of water[3]	0.015	0.017
Costs/m^3 of water on economic considerations	0.0187	0.020

Note: At official exchange rate: $1 = S£0.8.

[1] Estimates for the cost of windmill are based on the 5-m diameter prototype used in Sri Lanka and scaled accordingly.

[2] The pumpset includes a 16-hp engine and a centrifugal pump. The unit cost of water is very sensitive to the daily utilization rate. This is much more crucial for the diesel pumpset than for the windmill (since it does not depend on wind). Here it is assumed the diesel pumpset is used 4.5 hrs/day. Although this may be a reasonable assumption for farming conditions in the Sudan, if the number of hours and water output are doubled the cost/m^3 falls from $0.0175 to $0.0125. However, this utilization rate is likely to reduce the lifetime of the diesel engine, thus increasing unit costs.

[3] Cost per unit of volume of water during the first year has been estimated by using the following formula:

$$I + \sum_{n=1}^{N} O \frac{(1+p)^n}{(1+i)} = \sum_{n=1}^{N} V \times C \frac{(1+p)^n}{(1+i)}$$

where I = initial investment, N = lifetime, O = operation and maintenance per year, V = useful output per year, C = cost of water per m^3, p = price increase rate, n = index for the year, N = life of the project.

Source: Elaborated from Hamid and Jansen (1980).

The cost is $0.015/m^3 with a windmill and $0.0175/m^3 with a diesel pumpset. Since the diesel pumpset is utilized only 4.5 hrs/day on average, it can be argued that the same installation can be used to pump more water. If it pumps twice as much water, and if only fuel cost doubles, the cost per m^3 water from the diesel engine/centrifugal pump combination falls to $0.0125/m^3. However, if the diesel pumpset is used for about 9 hrs/day, then the expected lifetime of the pumpset is likely to be lower than the 7 years assumed in the analysis. This would raise the cost per m^3.

Apart from the user's cost of water pumping, economic costs of introducing windmills in the Sudan also need to be considered. Economic or real costs should exclude taxes, subsidies, and so on. According to the Petroleum General Administration, the market price of diesel in the Sudan reflects the real cost. For the economic evaluation, it is assumed that (i) the opportunity cost of capital is 10 percent (i.e., 1 percent higher than the prevailing interest rate); (ii) the tax on the import of diesel pumpset and construction materials is assumed at 20 percent; and (iii) the shadow exchange rate is at least 20 percent higher than the official rate. The two evaluations carried out (i.e., with and without a shadow price for foreign exchange) resulted in no difference in values for the windmill case and a difference on the order of less than 4 percent for the diesel pumpset. These calculations are ignored since the differences were small—most probably because the deduction of import taxes of 20 percent balanced the shadow price of foreign exchange (official rate plus 20 percent).

Two other effects that are important for the national economy are employment generation and foreign exchange savings. Since diesel pumpsets have been imported, the labor use is only for their sale and installation, which is estimated to be less than one worker-week for one pumpset. If a windmill is manufactured in the Sudan, it will create about 60 worker-days of work in the process of manufacture, sale, and installation, which is a lot more than that provided by imported diesel pumpsets.

Imported diesel pumpsets require approximately $6250 of foreign exchange per installation. For windmill systems, only basic construction materials need be imported; these are estimated at $1250 per set, leading to a direct foreign exchange saving of $5000 per installation. In addition, the use of a windmill saves about $406.20 in fuel imports per installation per year (at 1980 prices). Taking the present value over a period of 7 years (which is the lifetime of the diesel engine), the fuel savings are about $2000. One can therefore estimate that a windmill, if locally manufactured, can result in foreign exchange savings of about $7000 per installation over its lifetime.

Thus, in the Sudan, windmills seem more attractive to both users and society when compared with diesel pumpsets if the latter are used at an average rate of less than 5 hrs/day. However, the relative profitability of using a windmill is lower in cases where the alternative system (e.g., a diesel pumpset) could be used productively for a much greater number of hours.

OVERALL COMPARISON OF WINDMILLS WITH ALTERNATIVE TECHNOLOGIES

The country studies suggest that the financial unit costs of windmills are lower than those of the alternative technologies considered in the four countries, except in Sri Lanka (Table 5.8). At world prices (rather than local market prices) of kerosene and diesel, the windmills are found economically feasible in Sri Lanka and the Sudan. The unit costs of water pumped by windmills vary from one country to another and depend crucially on a number of locational and economic factors. For example, in Cape Verde the difference ranges from 2 percent in the ribeiras to 76 percent in the achadas, due to differences in wind speeds or pumping head. The cost

Table 5.8
Costs of Water Pumping for Irrigation with Windmills and Diesel/Kerosene Pumpsets in Four Countries ($/m³)

	Users' costs (at market fuel prices)		Economic costs (at world fuel prices)	
Country	windmill	kerosene/ diesel	windmill	kerosene/ diesel
Sri Lanka (1981)	0.053	0.05	0.053	0.071
Kenya (1981)				
Lower Tana (medium wind speed)	0.011	0.02	-	-
Kisumu (low wind speed)	0.050	0.066	-	-
Cape Verde (1980)				
Ribeiras	0.080	0.082	-	-
Sudan (1980)	0.015	0.0175	0.0187	0.02

Note: Figures for Sri Lanka, Cape Verde, and the Sudan are at 1980 prices and for Kenya at 1981 prices. "—" means not available, as economic analyses were not carried out in the original studies and estimates cannot be made on the basis of existing information.

Sources: Tables 5.1–5.7.

variation is about 34 percent in Sri Lanka under the assumptions of the economic costs of kerosene. In Kenya, the cost difference is 82 percent in favor of windmills in Lower Tana, which is the region of medium wind speeds, and about 33 percent in Kisumu, where wind speeds are relatively low. In the Sudan, the cost advantage to the windmills ranges from 7 to 16 percent, depending on the cost of diesel oil. This suggests that windmills can be financially and economically feasible in some regions and sites.

These findings are nevertheless affected by a number of assumptions that may not always reflect users' conditions, even in the locations under analysis. The most critical assumptions are about frequency and speed of winds, seasonal availability of groundwater, cropping pattern and farm size, and capacity utilization of windmills and of competing technologies. Hence the findings should not be regarded as final but, rather, as a reference for aiding decisions on investments and development of institutional capacities for evaluation of water pumping technologies and for R & D.

The above findings and their limitations bring out the following considerations about wind energy in developing countries:

1. The economics of a windmill is very location-specific, because output from a windmill depends crucially on wind speeds and other variables. In Kenya, the output from a windmill in Lower Tana is three times that in the Kisumu region and the unit costs ($\$/m^3$) in Lower Tana are almost one-fifth of those in the Kisumu area.

2. The water output and, hence, unit costs are sensitive to the total pumping head, as illustrated by the case of Cape Verde. In a ribeira (valley), with an average annual wind speed of 4.7 m/sec and an average total pumping head of 20 m, the annual water output was 16,000 m^3. With a higher (average annual) wind speed of 6.7 m/sec, the output was only 13,700 m^3 because the average total pumping head was 76 m in an achada (highland). This gives further evidence to the fact that the economics of windmills are site-specific, and that it is therefore necessary to survey sites that have potential for using wind energy economically.

3. The costs of a windmill are significantly affected by the cost of a storage tank. In Sri Lanka, the cost of the tank was $156 (about 15 percent of the total costs) and, in Kenya, it varied between $140 and $330 (depending on its size) and accounted for 10–25 percent of the total costs. In Cape Verde, it accounted for 60–65 percent of the unit costs of lifting water.

4. The unit cost also depends on assumptions made about capacity utilization of windmills and alternative devices. The total quantity of water available may not be utilized for a variety of reasons. For example, farm size may be smaller than necessary to fully utilize water output. Furthermore, there may be a mismatch between the monthly water requirement and output from a windmill on account of cropping patterns, wind speeds in different months, and so forth. Under these conditions, a comparison of unit

costs ($\$/m^3$) of water pumped can lead to erroneous conclusions. One method is to compare the unit costs for a given farm size and cropping pattern. For example, in Kenya it is shown that if water availability in the wells restricts the area irrigated to 0.5 ha, the windmill becomes an attractive proposition, unit costs being about 60 percent lower than those for a diesel pumpset. In this case, the capacity utilization of the windmill is only 38 percent whereas that of the diesel pumpset is even lower. Even where water is available, farm size may constrain capacity utilization. For small farms, it becomes important, therefore, to consider smaller sizes of windmills and diesel pumpsets as well as petrol and photovoltaic systems. The extent to which capital costs may or may not decrease with a reduction in farm size is crucial for the comparative economics of windmills and petrol, diesel, and kerosene pumpsets.

5. Due to considerable variation in wind speeds over different months, windmill capacity cannot be fully utilized. For example, in Sri Lanka (Jaffna district) the possible output from a windmill was 117 m^3/month in December–February, compared with 1405 m^3/month or higher during May–September. During October and November, the estimated output figures were 467 and 219 m^3/month, respectively. Given the monthly gross irrigation requirement of a given crop pattern, the output is inadequate even for a 0.1 ha farm during January–March, but sufficient for a 4 ha farm during May–September. Thus, the availability of water during the critical month determines the farm size that will be irrigated by a technology where water output varies significantly from month to month due to changes in resource availability. Also, in the case of Kenya the command area varies from 0.7 ha in December–January to 10.1 ha in September. Thus it was assumed that the command area of a 12PU500 windmill for a cropping pattern of rice, maize, and some other highland crops would be approximately 3.2 ha in the season of the long rains and 0.7 ha in the other season. Total annual water consumption is in this case 56,000 m^3/yr, which is 74 percent of the predicted windmill output. This shows that adjustments may have to be made in cropping pattern to improve the capacity utilization of a windmill. Thus the economics of a windmill also depends on whether or not implied changes in crop pattern are accepted by the farmers.

6. In some cases, it is assumed that a kerosene pumpset can be installed along with a windmill to supplement water output of a windmill during the off-season. For example, in the case of Sri Lanka (Jaffna), it is estimated that the use of a windmill together with a kerosene pumpset is economic for a 0.4-ha farm if the kerosene price is higher than Rs 5.02/1 ($0.312/1). In this case, the annual kerosene requirement is reduced by as much as 625 liters at an estimated economic cost of Rs 5.5/1. However, the same analysis shows that the use of a windmill together with a kerosene pumpset is not economic for a 0.2-ha farm, since the saving in kerosene is only 365 1/yr.

7. The economics of windmills is considerably improved if foreign exchange outflows and employment generation are taken into account. In Sri Lanka, annual foreign currency expenditures are estimated at $162 in the case of windmill and $292 in the case of a kerosene pumpset (i.e., a savings of $130 per year or a present rate of $650 per windmill over a 7 year period). In the Sudan, a locally produced windmill can save a total of $7000 in foreign exchange, at present values of expenditures in capital costs and fuels. In Cape Verde, the captial cost of a locally produced windmill is $1259, whereas the import cost of a diesel engine is $3800. Assuming that about 50 percent of the capital cost of a locally produced windmill is in foreign exchange, the direct savings in foreign exchange are about $3170 per installation. Adding to this a minimum saving in diesel oil of $320/yr, a total saving of $5000 (in present value) per installation is estimated. Also, in Kenya, a small windmill replacing a diesel pumpset is expected to provide annual foreign exchange savings of about $550/yr in capital cost installments and about $175 in fuel costs.

8. Use of windmills also creates additional employment, which is estimated as follows: in Sri Lanka, 16 worker-days of labor utilized just in the installation of a windmill; in Kenya, windmill production in 1982 employed more than 35 persons on a full-time basis; in Cape Verde, the extra employment created by a 5-m CWD prototype windmill is estimated at 550 worker-hours or about 70 worker-days; in the Sudan the estimate is that one windmill will create 60 worker-days of work in the process of manufacture, sale, and installation.

9. As expected, the economics of a windmill depend on the prices of fuels used in the analysis. In Sri Lanka, a windmill is not financially profitable compared with a kerosene pumpset if kerosene is priced at Rs 3.42/1 (implying a subsidy of 43 percent). However, when the economic price (shadow price or world price) of kerosene is taken (Rs 5.5/1), the unit cost of pumping with kerosene becomes higher (by about 33 percent) than the unit cost of a windmill. Similarly, in Kenya, the cost of a diesel pumpset goes up by about 2 percent when the shadow price of diesel is used instead of the market price. The cost of a windmill is reduced by about 12 percent when shadow prices instead of nominal prices (including import duties, sales taxes) are used.

SUMMARY OF CONCLUSIONS

The case studies suggest that windmills for irrigation pumping are financially and economically feasible under most conditions and assumptions described. However, the economics of windmills are very sensitive to water and wind patterns, farming conditions, fuel prices, and capacity utilization rates of both windmills and of competing options. Farm size and seasonal crop pattern are crucial factors, since they determine the utilization rates of the technologies under a specific seasonal pattern of water supply. Therefore, detailed information on these variables is required to ensure an adequate selection of locations for wind irrigation projects.

Given, on the one hand, the high capital costs of windmills and, on the other hand, the substantial effects of windmills on employment creation and foreign exchange savings, it is important to attribute explicit premiums to these benefits. If necessary, subsidies on investments should also reflect part of these benefits to society.

As in Sri Lanka, for example, subsidies on the prices of kerosene and diesel oil have made investments in windmills financially unfeasible, even though they were profitable from the viewpoint of society (taking the economic costs of those fuels into account at prices of 1980). In such conditions, it may be necessary to provide at least the same level of subsidies to investments in windmills to make them financially attractive. Changes in policies concerning fuel prices, taxes and import duties, and credit facilities to farmers may also be justified.

Evidently, the decline in world oil prices since 1981 will have reduced the economic attractiveness of windmills compared with oil-based pumpsets in the four countries reviewed. This, however, does not eliminate the usefulness of the case studies in providing, on the basis of past experience, basic parameters for ex-ante evaluation of similar projects. The next chapter will provide further analysis of such parameters with the application of an ex-ante evaluation of several alternative energy technologies for Northern India.

NOTES

1. The importance of monthly variation in water supply has been analyzed by Ghate (1980) and will be discussed in Chapter 6 in the economic comparison of windmills with other technologies for northern India.

2. It was not possible to evaluate the employment impact but, given that a windmill can be fully produced locally, it is likely to generate greater additional employment. In 1982, 16 worker-days of labor (mechanics and unskilled workers) were utilized just in the installation of a windmill.

6

Energy Alternatives for Irrigation Pumping: An Economic Analysis for Northern India

Ramesh Bhatia

Utilization of groundwater for irrigation has been steadily increasing in India since 1950/51. By 1980/81 the irrigated area had increased to 23 m.ha, which is about 57 percent of the ultimate potential. In northern states, the percentage of utilization has varied between 84 percent for Punjab and Haryana to 22–23 percent for Orissa and West Bengal. Estimates of additional area that could be irrigated through groundwater are: 3.2 m.ha in Uttar Pradesh, 2.0 m.ha in West Bengal, 2.5 m.ha in Bihar, and 1.1 m.ha in Orissa. Marginal and small farmers with farm holdings of less than 1 ha account for about two-thirds of the total holdings in Uttar Pradesh, Bihar, and West Bengal.

For more than a decade, policymakers have given major attention to electricity as a source of energy for irrigation pumping. Although in 1966 the numbers of diesel and electric pumpsets were approximately equal (0.5 million), there were about 5.2 million electric pumpsets as compared with 3.0 million diesel pumpsets in March 1985. The number of electric pumpsets has increased faster because of concerted efforts in rural electrification to promote groundwater utilization. Despite massive investments,[1] the benefits from rural electrification have not been commensurate with costs. As a result, other alternatives are increasingly being considered.

Irrigation systems using the following renewable energy technologies are being field tested: solar-photovoltaics and solar-thermal devices, windmills, and multi-fuel pumpsets using diesel with biogas, pyrolysis gas, or ethanol. About 200 photovoltaic systems and 400 windmills have been installed in India during the last few years. However, the development and use of renewable energy sources for irrigation has been considerably affected by the government's policy of subsidizing the use of electricity and diesel oil in the past. Given the low prices of conventional energy sources, the use of renewable sources has not been found financially attractive by farmers.

The objectives of this paper are to provide:

- a brief review of the field experience with renewable energy technologies used for irrigation water pumping in India and abroad;
- A methodology for economic evaluation of alternative energy sources (for given quantities of water) with the help of a case study of the Ghazipur area in the Gangetic plains of northern India; and
- some policy guidelines for pricing and organizational aspects of promoting renewable energy technologies.

A REVIEW OF FIELD EXPERIENCE WITH RENEWABLE ENERGY TECHNOLOGIES

Solar Photovoltaic Systems

Photovoltaic water pumping systems consist of an array of solar cells that are oriented to face solar radiation through manual adjustments or with the help of automatic trackers. In some cases, solar radiation may be concentrated and then focused on the array. The electrical output of the array energizes a motor, generally a direct current (DC) type, which in turn operates a pump through a mechanical transmission. In some systems, batteries are used as buffers between the array and the motor. Some of the systems have been provided with power conditioning devices that try to maintain system operation at peak efficiency by matching the load with the output of the arrays.[2]

A fairly exhaustive analysis of small-scale solar-powered pumping systems has been carried out by William Halcrow and Partners for the UNDP/ World Bank (Halcrow et al., 1983). As many as 12 commercially available systems, ranging in array power from 200 to 1260 peak watts (W_p), were field tested in Mali, the Sudan, and the Philippines. Among the units that ran without trouble were Pompes Guinard and Arco Solar, whose overall peak efficiency was about 3.5 percent (at 1000 W/m^2) in terms of gross cell area and 2.7 percent in terms of overall array area. These field tests also provided estimates of daily efficiencies based on measured hydraulic output and solar radiation input. This was about 2.1–2.3 percent in terms of gross cell area and 1.3–1.8 percent in terms of array area on a standard clear day (6 kWh/m^2 daily).

The daily efficiency depends on the level of solar radiation. The field data on a fairly successful unit (149 days of trouble-free operation in Mali) were used to estimate the relationship between daily overall efficiency and daily global radiation. The peak daily efficiency in terms of array area was about 2.1 percent, whereas peak instantaneous efficiency was about 3.4 percent. In our analysis, for the reference case, a figure of 3.4 percent peak instantaneous efficiency is used. A case in which peak instantaneous efficiency of 4.6 percent is achievable is also included to reflect the effect of technological advances on future efficiencies.

Results of field trials of 17 photovoltaic systems in Pakistan have been reported in Bhalla, James, and Stevens (1984). The area commanded by a system of 250 W capacity (under 100 mW/cm^2 insulation) ranged from 1.0 to 3.6 ha while the average area irrigated per day varied between 0.1 and 0.2 ha. The study describes the type of problems that these units experienced in the field, which included failure of motor and/or pump in 7 out of 12 systems. No fault was detected on the solar panels themselves, which were handled by the farmers with utmost care. The study points out that it would be difficult to reach a firm conclusion about the viability of photovoltaic systems in Pakistan and that price reductions would be required if they were to compete with existing options.

Under the Solar Photovoltaic Demonstration Programme, the government of India has installed some 200 photovoltaic systems and about 130 of them are said to be for irrigation. In 1985, the performance data for these installations were being collected but had not been analyzed and reported. Two of these field units at Navsari (Gujarat) and Kotputli (Rajasthan) were visited to collect some information. At Navsari, a 300 W_p system given by the Central Electronics Limited was operating during 1983/84 to irrigate a plot of 0.45 ha under wheat and 0.45 ha under cotton. The quantity of water pumped (at 1–3.4 m head) ranged from 518 m^3 in January to 1150 m^3 in April (Table 6.1). At Kotputli, the water output was estimated at 800 m^3 during March. An evaluation is available in respect of one photovoltaic system imported from Solar Electric International (SEI) and tested at Water Technology Centre of Indian Agricultural Research Institute (IARI), New Delhi, during 1980. The overall efficiency of the system was reported up to a maximum of 3.44 percent (at 820 W/m^2). The maximum efficiency was 9.57 percent for the array and the Maximum Power Point Tracker (MPPT), and 35.9 percent for the motor and pump combined.[3] At low heads of 3–5 m, the estimated monthly water output ranged from 600 m^3 in January to 714 m^3 in March to provide irrigation to 0.6 ha of wheat. The figures for SEI systems in Orissa range from 600 m^3 to 1400 m^3 of water output per month (Table 6.1).

Table 6.1
Field Data on Output of Photovoltaic Systems for Irrigation Pumping in India

Institution and specification	Monthly water output (m^3) January	February	March	April	Irrigated area (ha)
IARI, New Delhi[1] (SEI, 240 W_p)	601 (3-5 m)[2]	602 (3-5 m)[2]	714 (3-5 m)[2]	-	0.6 (wheat)
GAU, Navsari[3] (CEL, 300 W_p)	518 (1-3.2)[2]	893 (1.1-3)[2]	930 (1.8-3.2)[2]	1,150 (1.8-3.4)[2]	0.45 (wheat)
KRECS, Kotputli[3] (CEL, 300 W_p)	-	-	803 (2.8 m)[2]	-	0.45 (cotton)

	Days of operation	Daily water output (m^3)	Head (m)	Irrigated area (ha)
Orissa[4] (SEI, 222 W_p)				
Bhubaneshwar[5] (A)	30	32.2	3.0	0.41
" (B)	91	24.5	3.6	1.82
Astarang	54	50.6	3.0	2.02
Pratapnagari	23	19.3	3.6	0.60

Note: "—" means unavailable or unreliable data. The data for systems installed at Orissa cannot be presented like the data for other systems. See sources below for further technical details.

[1]System involves 1 m × 2 m arrays, MPPT, and submersible pump. Daily average figures from IARI (1980) are converted here into monthly figures. A total quantity of 1098 m^3 of water was actually delivered to the wheat crop from 16 January to 2 April 1980.

[2]Indicates head in meters.

[3]Personal communication during visits to the site. These are CEL demonstration units of 20 modules, each module comprising 36 cells of 76 mm diameter, DC motor, and centrifugal pump.

[4]Charlu and Dutt (1982).

[5]Includes two cases, A and B.

Source: See notes.

Solar-Thermal Systems

Although a number of experimental solar-thermal prototypes are being developed in India, day-to-day performance data are not available. However, the following figures for instantaneous efficiencies of overall systems have been quoted: 0.16–0.35 percent for Birla Institute of Technology (BIS), Pilani, 0.17–0.57 percent for Central Salt and Martine Chemicals

Research Institute (CSMCRI), Bhavnagar, and 0.45 percent for Jyoti Limited, Baroda. There are no reports of trouble-free operation of any system for a reasonable period to assess the performance of these units in terms of output of water pumped.[4]

According to Halcrow et al. (1983), one prototype from Solar Pump Corporation was installed in the Sudan, but continous operation was not achieved due to a variety of problems. During the same project, four solar thermal systems were tested at the manufacturers' premises and their instantaneous efficiencies were measured. A Dornier system showed efficiency of 3.5 percent (engine and pump) when operated with an electrical heater (poor sunshine during the period),[5] but its overall efficiency has been estimated at 1.9 percent. The results of a mathematical modelling exercise have been reported in Halcrow et al. (1983). The daily efficiency considered in the study regarding systems based on nontracking flat plate collectors is 1.0 percent, but instantaneous efficiency can be higher. This seems consistent with the efficiency of 1.9 percent reported for the Dornier system, bearing in mind that this system is based on a more efficient heat-pipe collector. Efficiencies are better and total system costs are lower in cases where tracking is provided by flat plate collectors and with the use of concentrators.

Biogas Systems

Use of biogas (*gobargas*, or methane) for irrigation pumping has been considered in several studies.[6] It is often assumed that engines using gas, petrol, or diesel can be easily and cheaply adapted to the use of biogas. In the case of spark-ignition engines, biogas can replace petrol completely, but in compression-ignition engines, only about 80 or 90 percent of diesel can be replaced.[7]

In a comparison of energy alternatives for tubewell irrigation pumping for small farmers of Eastern Gangetic Plains, Bhatia (1980a) concluded that at capital and operating costs of 1979, biogas (from Janata design biogestors) was the most economic alternative from the viewpoint of society.[8] From the farmers' viewpoint, subsidized electricity from the regional grid was the cheapest alternative. A petrol engine using 100 percent biogas from a conventional biogas plant (KVIC design)[9] was financially more attractive than a diesel engine as well as a dual-fuel engine using 20 percent diesel plus 80 percent biogas from a KVIC plant. For the dual-fuel case, however, if capital costs were reduced by about 50 percent (as claimed for the Janata design popularized by the Uttar Pradesh state government), dual-fuel engines were more attractive to farmers than diesel engines.

A major manufacturer of diesel engines (Kirloskars) claimed that a number of dual-fuel engines (80 percent biogas plus 20 percent diesel) have been sold in the market in the last few years. However, no list of buyers could be made available and, therefore, no farmer who was actually using biogas in a dual-fuel or a petrol engine could be interviewed. As reported in Ghate (1980) and Bhatia and Niamir (1979), biogas was supposed to be used in a dual-fuel engine to operate a tubewell, a flour mill, a power thresher, and a chaff-cutter, as part of a community biogas project at village Fatehasingh-ka-Purwa in Uttar Pradesh. In a field visit to that village in 1979, it was found that the tubewell had not been operated for more than a few hours due to various difficulties. However, the dual-fuel engine had been used to operate a flour mill and power thresher for almost 600 hours between 1979/80 and 1981/82. For operating a 5 hp dual-fuel engine for one hour, about 0.55 liters of diesel plus 2.22 m^3 of biogas are required. This level of fuel consumption is slightly higher than that assumed in the two studies mentioned above, namely 2.14 m^3/hr—that is, 75 ft^3/hr or 15 ft^3/hp-hr (horsepower-hour)—plus 0.11 1/hp-hr of diesel, also for a 5 hp engine.

Data on another experience with biogas in a dual-fuel engine for irrigation have been obtained from PSG College of Technology, Coimbatore. The objective of this project (over the 1979/81 period) was to study the possibility of using biogas in diesel and spark-ignition engines and the long-term effect of this use on engine components. A report on the project claims that the technology is already mature. They have a gas storage and production capacity of 150 ft^3 (approximately 4 m^3), and the plant requires 50 kg of fresh dung per day. They use a mix of 30 percent diesel and 70 percent biogas, which gives an estimated requirement of 0.3 kg (0.36 liters) of diesel and 1,500 1/hr of biogas for a 5 hp (3.65 kW) dual-fuel engine. This gives figures of biogas and diesel consumption of 0.28 m^3/hp-hr and 0.072 1/hp-hr, respectively. In our analysis, it is assumed that under field conditions biogas and diesel consumption will be somewhat higher (i.e., 0.368 m^3/hp-hr of biogas plus 0.135 1/hp-hr of diesel).

Also in the Indian Institute of Technology, Bombay, a 28.3 m^3 biogas plant (KVIC design) was used during 1976–1985 to study the technical performance of dual-fuel engines. They found that a biogas-cum-diesel engine operates well, so long as the biogas supply is available; the moment biogas runs out, the engine automatically shifts to pure diesel operation. This provides the necessary flexibility to the system, since it can operate on diesel alone.

In summary, it seems that biogas is technologically suitable for irrigation, although it has not been financially attractive because of subsidies on prices of electricity and diesel. While petrol engines can run on 100 percent biogas, they have not been used in irrigation because the price of petrol is much higher than that of diesel. Furthermore, a high price of petrol (due to high excise duty) has resulted in non-use of petrol engines for irrigation. If appropriate pricing and capital subsidy mechanisms are adopted, the use of biogas for irrigation can be made profitable for the average farmer.

In our calculations, it is assumed that a mixture of 70 percent biogas plus 30 percent diesel is technically feasible in the case of a dual-fuel engine. However, the optimal mix of biogas and diesel depends on the following factors:

i) availability of dual-fuel engines and petrol engines and their costs;
ii) capital costs of biogas plants of different sizes and designs;
iii) reduction in output of biogas in winter months; and
iv) the pattern of irrigation water requirements in different months reflecting crop rotations, rainfall conditions and so on. These aspects are discussed later in this chapter.

Gasifiers

Gasification of biomass is considered to have great potential for water pumping in developing countries. The technology is nearly a century old. The theoretical efficiency of a typical gasifier is approximately 70 percent and even a simple design can reach an efficiency of 60 percent under normal operation. Some studies have recently appeared on the techno-economic aspects of gasifiers in developing countries, especially India. Notable among these are: Hurst (1983); Hurst and Rogers (1982); Rogers (1981); Joshi, Seckler, and Jain (1983); de Lucia and Lesser (1980); Datta and Dutt (1981); and Foley and Barnard (1983). According to Hurst and Rogers, it is advisable to use charcoal as a fuel for small systems. Other fuels (wood and agricultural residues) produce tars that are harmful to engines and can only be removed by expensive filters. Charcoal manufacture complicates the system and reduces the overall efficiency by approximately 50 percent. Pessimistically, the overall efficiency of a system involving charcoal manufacture, gasifier and diesel engine (or gas engine) is 5 percent. However, in a cost/benefit analysis, Rogers (1981) has shown that use of charcoal (instead of wood) is the cheapest alternative for a micro-gasifier suitable for use with a 2.25 kW (3 hp) engine.

In the Philippines, a number of gasifiers had been built and used in irrigation pumps, jeepneys, tractors, and so on. However, no report on the field performance of these units was available by early 1984. It was reported that the saving in diesel was 50–70 percent if a gasifier of 20 hp (or more) was used. In the case of small gasifiers (5–10 hp), the diesel saving was expected to be only around 30 percent. The fuel used was charcoal from fuelwood and from coconut-shell. Direct use of fuelwood was not tried because it was expected to cause technical difficulties.

Joshi, Seckler, and Jain (1983) report on the techno-economic aspects of producer gasifiers using fuelwood. According to them, a 8.8 kW, 1500 rpm dual-fuel engine with gasifier was successfully run for over 400 hours at Mogar in Gujarat. The engine at the demonstration site was dismantled after 325 hours of operation to check the condition of various components. That study, however, made no observation on engine performance that can be specifically attributed to the use of gas. The same study reported that a gasifier powering a 5-hp engine ran satisfactorily for about 100 hours. The gasifier used firewood directly, and it was estimated that approximately 3.5 kg of firewood replaced one liter of diesel. In the experiment, 76 percent of diesel was replaced by gas in a dual-fuel engine. In the economics worked out in that study, the benefit/cost ratio was calculated as 3.1 for a gasifier used to irrigate 90 ha on a rotation of paddy and wheat.

Windmills

In India, deliberate efforts were made to design and utilize windmills for irrigation in small farms under the TOOL-ORP Project of Ghazipur. The windmill developed—the 12PU500, also known as the Ghazipur windmill—is comparable in size to the WP-2 of the National Aeronautics Laboratory (NAL), both having a rotor diameter of 5 m, but differs in terms of level of technology. The WP-2 is a conventional design requiring high-quality components and the Ghazipur windmill is characterized as an "appropriate" technology. The latter is somewhat cheaper but seems to require more frequent maintenance. Attempts have been made at NAL to design a larger windmill (MP-1) using a sail rotor (7.5 m diameter) and rotary pumps. The basic approach has been to optimize the design of this windmill for low-wind months, which require a larger rotor diameter. The choice of rotary pumps has been made to improve the reliability in handling irrigation water. There have also been several other prototype-development activities such as a yacht-sail windmill at SWRC, Tilionia (near Ajmer), a sail windmill at Gaudagaon (near Solapur), and Thai sail-type windmills at Yellamanchilli (near Vijayawada), but they have not been sufficiently developed and documented. The other developments in "appropriate technology" are: a sail-type Savonius rotor developed by the Indian Institute of Science, Bangalore, and a sail-type windmill made entirely from locally available materials developed by Murugappa Chettiar Research Center at Madras—both are relatively small.

According to the Department of Non-Conventional Energy Sources of the government of India, more than 400 water-pumping windmills (mainly of the Ghazipur type) have been set up in the country under the field trial/demonstration program. Monitoring and assessment of the socioeconomic impact and technical performance of these windmills was under way in early 1985. Until 1983, the only sources of performance data for India were the TOOL-ORP Evaluation Report (December 1980) for Ghazipur and a

Table 6.2
Measured, Predicted, and Estimated Monthly Water Output of a Windmill in Ghazipur (m^3/month)

Month	Mean wind speed[1] (km/hr)	Measured[2] (Head 6 m)	Predicted[3] (Head 6 m)	Estimated[4] (Head 5 m)
January	6.9	118	850	450
February	7.0	523	950	430
March	8.7	915	1,850	910
April	9.5	874	2,600	1,140
May	10.5	1,922	3,150	1,590
June	10.4	1,794	3,350	1,500
July	9.1	1,381	2,200	1,040
August	9.1	1,751	1,900	1,040
September	8.5	875	1,950	820
October	6.7	324	800	410
November	5.5	359	400	220
December	6.6	143	500	400

Note: Data refer to 12PU500 Ghazipur/Allahabad windmill.
[1]Data from TOOL-ORP (1980, Table 4.1).
[2]The data from Nov. to July are for one windmill and Aug. to Oct. for another. These data are based on actual measurements of water output as reported in Ghate (1980).
[3]This estimate is based on data in TOOL-ORP (1980, Table 4.1). The short-term test data were converted into monthly output using frequency distribution of wind speeds. The water output under the column "predicted" is assumed to be equal to 50 percent of the output estimated from short-term test data.
[4]Based on 10 percent overall average efficiency of the windmill. See SDCA (1985).
Source: See notes.

study by Ghate (1980).[10] Some more recent data were made available from field visits to Ghazipur villages as part of a study sponsored by the Swiss Development Cooperation Agency (SDCA, 1985). Under the Ghazipur project, one windmill was monitored from day to day at the Rural Technology Institute (RTI) between December 1978 and July 1979 and another at Rasadi between August and October 1979. The data on monthly output and mean wind speed[11] can be used to calculate overall monthly efficiencies ranging between 20 percent and 28 percent except for December 1982, when it seems to have been 15 percent. These are exceptionally high values not recorded anywhere else in India. During December 1982, a day-to-day comparative monitoring of a Ghazipur-type windmill installed at the Central Power Research Institute (CPRI), Bangalore, showed that efficiency was about 2 percent or less. The field survey by the SDCA team showed that 11 out of 12 farmers using ORP-TOOL windmills in Ghazipur had to purchase water during the *Rabi* (winter) season[12] (SDCA, 1985). Although it was not possible to obtain data on the quantity of water actually purchased, the estimates made from data on the number of hours of pumping of diesel/electric pumpsets to supplement the water requirement showed that the Ghazipur windmill was not capable of irrigating more than 0.4 ha (1 acre) during the Rabi season with the wind speeds prevailing in that area.

Given the inadequacy of data on performance of multivane windmills such as the Ghazipur type, a tentative value has been estimated as follows: based on the measured efficiency of a MP-1 windmill, which is about 5 percent, one can expect a multivane rotor coupled to a piston pump to have an overall efficiency of 10 percent. This higher value is mainly due to piston pump efficiencies, which are almost twice those of centrifugal pumps, and partly due to a clean rotor compared to a trapezoidal sail-type in the MP-1 windmill. The assumed efficiency of 10 percent is a compromise between the estimated values from performance analysis at Ghazipur (20 percent) and CPRI, Bangalore (2 percent).

One important aspect of using windmills in regions such as Ghazipur has been pointed out by Ghate (1980). For two windmills analyzed, one-half of the total water output is available in four months: March to June (see Table 6.2). Thus, windmill irrigation benefits are heavily dependent on the utilization of water during the *Zaid* (summer) season (in between the main crop seasons of the Rabi and the Monsoon). The Rabi season utilizes the most irrigation, at present, but has the leanest wind regime. (The months of December and January account for 1–1.5 percent of total annual output.) Even in the *Kharif* (monsoon) there is not enough wind to satisfy the irrigation requirements of a paddy crop during rainless spells. Thus, to meet the water requirements of Rabi crops, the number (or size) of windmills will have to be increased to make this option comparable with other alternatives.

ECONOMICS OF ENERGY ALTERNATIVES

An Overview of the Methodology

The economic evaluation carried out in this section is for conditions in the eastern Gangetic plains represented by Ghazipur, Uttar Pradesh. Two farm sizes are considered: 0.4 ha and 1.0 ha. The water pumping options studied are:

i. electric pumpsets using electricity from the regional grid;
ii. diesel pumpsets;
iii. dual-fuel engine using biogas or pyrolysis gas (from a gasifier);
iv. solar-photovoltaic and solar-thermal systems; and
v. windmills.

The economic evaluation and ranking of technologies has been attempted in terms of cost-effectiveness criteria for "equivalent" irrigation systems. In this methodology, each individual technology is designed so as to provide the same quantity and quality (time pattern) of water for irrigation, thus giving the same level of agricultural benefits. These systems are compared and ranked in terms of the sum of present value of capital costs and operating costs incurred over a fixed period (e.g., 10 years). Since the life of each component of capital equipment may vary from one technology to another, this would result in inclusion of replacement costs of some types of capital equipment (e.g., a gas holder may be replaced every 10 years or a petrol engine may be replaced after 5000 hours of use). As benefits from irrigation are the same (since the systems are equivalent), the alternative with the lowest present value of total costs is considered the best (i.e., giving the highest benefit/cost ratio).[13]

The methodological steps used to carry out the economic evaluation in this chapter have been listed in the introduction to Part II (Chapter 1) in order to show the methodological differences between this chapter and the two previous chapters. For ready reference, these steps are recapitulated below:

i. determine a few representative cropping patterns and estimate monthly water requirements for each crop rotation on the basis of agro-climatic conditions and effective rainfall;
ii. analyze resource availability in the context of solar and wind energy, in particular the magnitude and distribution of solar radiation and wind speed;
iii. identify critical months for each crop rotation—that is, months with highest ratios between water requirement and gross solar radiation (or mean wind speed);
iv. estimate the installed capacities of equipment to deliver estimated irrigation water based on an analysis of critical months and relevant energy resource data;
v. estimate the capital costs and operating costs of each system from the viewpoint of society using shadow prices for major inputs; and
vi. estimate the present value of the sum of these costs (using an appropriate SRD) for each system for the purpose of comparison and ranking.

These steps are described in the subsequent subsections, as applied to the case of three crop rotations in Ghazipur.[14] Although the basic analysis has been carried out to reflect the viewpoint of society, the financial feasibility of each system has also been considered along with a discussion of fiscal policies necessary for the promotion of renewable sources of energy.

Estimation of Irrigation Water Requirements

A procedure for estimating crop water requirements has been described in FAO (1977).[15] This involves determining the evapo-transpiration rate (ET_0) for a reference region on the basis of climatic data such as temperature, humidity, sunshine, radiation, and so on. The effect of crop characteristics on crop water requirements is given by $ET_{crop} = Kc\ (ET_0)$, where ET_{crop} is the evapo-transpiration for the crop and Kc is the crop coefficient (based on agricultural practices, size of farm, method of irrigation, etc.).

To estimate irrigation water requirement (i.e., volume of water that has to be pumped), one uses evapo-transpiration for a crop (ET_{crop}), effective rainfall (ER), and field application efficiency (E_a). Since irrigation is to be provided from wells located within the farm, conveyance losses are not relevant but water-course losses and evaporation losses are important. These are reflected by E_a, which is the ratio between water available directly to the crop and that delivered by the pump (or storage tank). It is reported (*FAO Bulletin,* 1977/80) that in irrigation practices, such as graded borders, furrow, corrugation, and so on, one may assume a conservative figure of 0.55 (for non-rice crops). This is assumed to include water-course losses where there are no lined channels in the field. Furthermore, this factor is assumed to be valid between 0.4 and 2.0 ha. However, for rice this factor is still lower, 0.32, due to higher percolation losses.

An overestimation of irrigation water requirements can result in the installation of over-capacity for pumping. To avoid this, it is necessary to assess the extent of tolerance of various crops, in the event that available moisture is less than the saturation level of the soil. For instance, in the case of wheat, it is reported that, even by maintaining a ratio of 0.75 between irrigation water and cumulative-pan evaporation,[16] the yield is hardly affected irrespective of the stage of crop growth. A lower ratio of 0.6 can also be maintained, ensuring proper moisture availability at critical growth stages. However, it is assumed that for the critical month for which a particular technology is designed, water requirement can be reduced up to 75 percent for a non-rice crop (and not reduced for rice). The calculated water requirements (modified by a factor of 0.75) at the field inlet (or pump/tank outlet) for three cropping patterns (rotations) are given in Table 6.3 for the selected region in m^3/ha/month.

Assessment of Solar and Wind Resources

The relevant data for photovoltaic arrays or flat plate solar thermal collectors are global radiation, comprising both direct and diffused radiation. The network of radiation stations includes 16 principal and ordinary stations

Table 6.3
Irrigation Water Requirements (Iw) for Three Crop Rotations in Ghazipur Region (m^3/ha/month)

	Crop rotation I			Crop rotation II			Crop rotation III		
	Wheat + rice			Wheat + bajra			Sugar-cane		
Month	ET_{crop}	Effective rainfall	Iw*	ET_{crop}	Effective rainfall	Iw*	ET_{crop}	Effective rainfall	Iw*
January	811	250	760	811	250	760	792	250	740
February	1,235	219	1,390	1,235	219	1,390	703	219	660
March	509	119	530	509	119	530	815	119	950
April	-	39	-	-	39	-	1,902	39	2,540
May	-	123	-	-	123	-	3,305	123	4,340
June	-	919	-	-	919	-	2,715	919	2,450
July	2,496	(2,140)	1,110	1,021	(2,140)	-	2,382	(2,140)	330
August	1,876	(2,400)	-	1,637	(2,400)	-	1,790	(2,400)	-
September	1,698	(1,580)	370	1,617	(1,580)	50	1,697	(1,580)	160
October	1,589	595	3,110	771	595	240	1,585	595	1,350
November	1,110	96	3,170	669	96	780	1,225	96	1,540
December	519	33	660	519	33	660	957	33	1,260

Note: For all crops except rice, irrigation water requirement (Iw) for most critical month was reduced to 75 percent of the theoretical requirement. Thus, for non-rice crops, Iw* = 0.75 Iw. For rice Iw* = Iw. Iw is given by the formula: $Iw = (ET_{crop} - ER)/E_a$, where ET_{crop} is evapo-transpiration rate of the crop; ER is effective rainfall; E_a is field application efficiency: 0.55 for non-rice crop and 0.32 for rice. Thus, for example, for sugar cane, in January, $Iw = 0.75\ (792 - 250)/0.55 = 740$. Effective rainfall was estimated at 70 percent of effectiveness factor (whenever normal rainfall is higher than 100 mm/day).

Source: SDCA (1985).

where recordings of global radiation are made. The station nearest to Ghazipur, for which calculated values have been published (Mani and Rangarajan, 1981), is Varanasi. Mean values of daily global radiation for each month, in kWh/m^2, are given in Table 6.4. The monthly mean wind speeds for the location of interest (i.e., Ghazipur) are given in Table 6.2.

Estimation of Capacities

Capacities of photovoltaic pumping systems

Given the estimated water requirement and data on solar radiation, the next step is to estimate the installed capacity to provide this water over

Table 6.4
Mean Value of Daily Global Radiation for Ghazipur Region (kWh/m²)

Month	Mean value
January	4.54
February	5.41
March	6.19
April	6.92
May	7.20
June	6.37
July	5.31
August	5.04
September	5.41
October	5.63
November	4.84
December	4.38

Source: Mani and Rangarajan (1981).

monthly periods. For each crop rotation, critical months are first identified as those which show the highest ratio between monthly water requirement and the monthly gross solar radiation. In cases where rice happens to be one of the crops, this ratio is optimized with respect to the peak ratio for non-rice crops under the same crop rotation.[17] For the crop rotations considered for Ghazipur, the critical months are February for crop rotation I (rice and wheat) and crop rotation II (bajra and wheat). The critical month estimated for crop rotation III (planted sugar cane) is May. Estimates are made for installed capacities to match energy requirements in the critical month of each crop rotation (Table 6.5). The estimated requirement of hydraulic output is calculated from the water requirement for the critical month (in m^3 as given in Table 6.3) multiplied by the total head[18] (in m) and a conversion factor of 2.724×10^{-3} to arrive at kWh units of work.[19] Estimates of average array-plane global daily radiation during the critical months of February and May are given for Ghazipur (Table 6.5, row 3). The tilt is selected in discrete positions of latitude ($\pm$ 15°) to maximize energy availability during the critical month. The hydraulic output from 1.0 m^2 of array is then calculated in terms of kWh per month. This is arrived at by multiplying the figure for average daily radiation (7.13 kWh) by the corresponding overall daily efficiency (2.12 percent), which can be read in Figure 6.1, and finally multiplied by the number of days in the month. The gross cell area required (in m^2) is estimated by dividing the total requirement of hydraulic output by the hydraulic output from 1.0 m^2 of array.

With the reported array efficiency of 9.2 percent in terms of gross cell area, the peak ratings (in 1000 W/m^2 of sunshine) are determined. As given in Table 6.5, a photovoltaic system of around 400 W_p will be sufficient to

Table 6.5
Estimation of Required Capacity (in Wp) of Photovoltaic Systems in Ghazipur

	Crop rotation		
	I	II	III
1. Critical month	February	February	May
2. Assumed tilt[1]	lat + 15°	lat + 15°	lat - 15°
3. Average array plane global radiation during critical month (kWh/m² daily)	7.13	7.13	7.03
4. Calculated hydraulic output[2] from 1 m² of array (kWh/month)	4.232	4.232	4.598
5. Requirement of hydraulic output for the field[3] (kWh/month)	18.93	18.93	59.11
6. Gross cell area required (m²)[2,4]	4.47	4.47	12.86
7. Required capacity (Wp rating)[5]	410	410	1,180

Note: The total head is assumed at 5 m. See Table 6.3 for information about crop rotations, water requirements and critical month.

[1] The tilt is selected in discrete positions of latitude to maximize energy availability during the critical month.

[2] This is calculated by multiplying the figure of average daily radiation (row 3) by efficiency, read from Figure 6.1, corresponding to this value of daily radiation, and finally multiplied by number of days during the month (e.g., for February, 7.13 × 0.0212 × 28 = 4.232).

[3] This is estimated from water requirement (Table 6.3) for the critical month (in cubic meters) multiplied by the total head in meters and a conversion factor of 2.724×10^{-3} to arrive at units of work in kWh.

[4] Obtained by dividing the value in row 5 by the value in row 4.

[5] Assumed array efficiency of 9.2 percent in terms of gross cell area and 1000 W/m² for peak solar radiation as reference.

Source: Notes and text.

provide irrigation to a 1-ha farm under crop rotations I and II. However, sugar cane will require 1180 W_p capacity. The size of motor and pump are appropriately selected to maximize the utilization of solar energy.

Capacity of Solar-Thermal System

So far, in Indian systems, daily efficiencies of about 0.5 percent have been demonstrated. This may be treated as the daily average peak value to correspond to the same distribution as in Figure 6.1. This means that the collector area has to be increased by a factor of 4.2 over what has been estimated as the gross cell area. It may be noted that this technology is being treated in

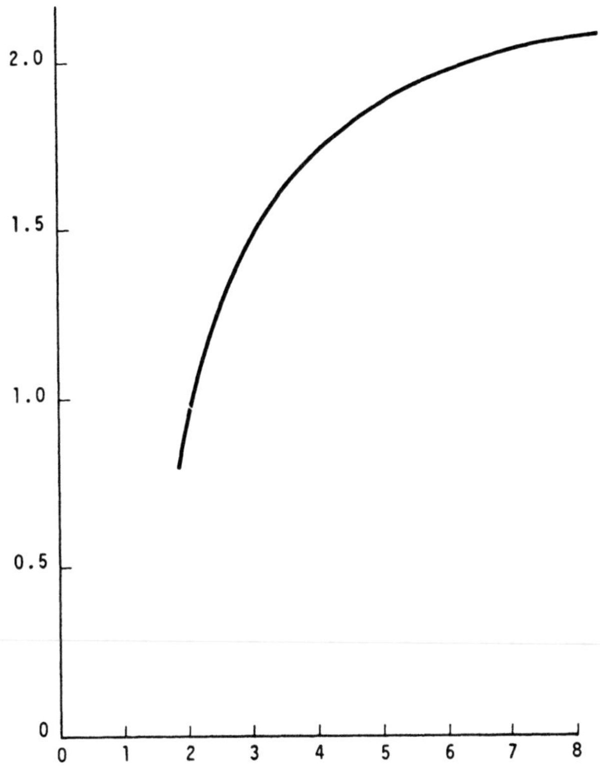

Figure 6.1
Assumed Overall Daily Efficiency Distribution for a Photovoltaic Pump
Global daily radiation in the plane of the array (kWh/m²)

Source: Elaborated from data in Halcrow et al. (1981), and reported to SDCA (1985).

a somewhat optimistic manner. Such systems have significant thermal inertia for both start-up and shut-down. On days of interrupted sunshine, their performance may not be as good as that of photovoltaic systems. Besides, these would require much more than 220 W/m^2 for the start-up.

We may also include an eventual possibility of achieving peak daily efficiency (defined at 6 kWh/m^2 of daily insolation) of 1.0 percent, which would then mean a reduction of collector area by a factor of two from what is considered necessary in current estimates.

Capacities of electric motor, diesel engine, and dual-fuel engine

Proper selection, installation, operation, and maintenance of centrifugal pumps and related prime movers (diesel engine or electric motor) are very important aspects to ensure their economical, long and trouble-free service. Major considerations for selecting a pumpset are: (a) discharge to be pumped, (b) total head, and (c) range of variation of water level over which the pump has to operate (or operating conditions). Given the estimated discharge and the total head over which water is to be lifted, the pump has to be selected so that it operates at maximum efficiency within the expected range of water level fluctuations on the suction side.

For Ghazipur, a 5 m head has been assumed. For each of the crop rotations and farm sizes (0.4 and 1.0), the required water horsepower has been estimated using the standard formula. The brake horsepower (bhp) has been estimated assuming a combined efficiency of 30 percent[20] for the pump and transmission. To take care of seasonal variation in load, however, the selected bhp of electric motors should be higher than the estimated value.[21]

Using these assumptions, we have calculated the required bhp of the diesel engines and electric motors for meeting the pumping requirements of different crop rotations in the region.[22] The same figures have been used for dual-fuel engines using biogas or pyrolysis gas. The details of estimated capacities are given in Table 6.6.

Table 6.6
Estimation of Required Equipment Capacity (hp) for Conventional Systems in Ghazipur

Crop rotation		Critical month	Gross irrigation (water requirement) m^3/month/ha	Required discharge[1] l/sec	Required discharge[1] m^3/hour	whp[2]	bhp[3]	Engine bhp[4]
I	Wheat + rice	November	3,170	3.67	13.212	0.24	0.80	1.20
II	Wheat + bajra	February	1,390	1.60	5.792	0.11	0.36	0.55
III	Sugar-cane	May	4,340	5.02	18.072	0.33	1.10	1.65

Note: Assuming that the number of hours of pumping during critical month is 8 hrs/day; farm size is 1.0 ha, total head is 5 m, and combined efficiency of pump and transmission is 0.3.

[1]Discharge (1/sec) = Water requirement (in m^3/month/ha) × 1,000 (in liters) divided by number of hours of pumping during critical month [hrs/day × 60 (min) × 60 (sec) × number of days in month].

[2]Water horsepower (whp) = (Discharge × total head)/76.

[3]Break horsepower (bhp) = whp divided by the combined efficiency of pump and transmission. For example, for Ghazipur, for crop rotation I, bhp = 0.24/0.3 = 0.8.

[4]Engine bhp = 1.5 × (bhp) up to 2 hp; 1.3 × (bhp) between 2 to 3, and 1.2 × (bhp) when bhp is greater than 3.

Source: SDCA (1985).

Capacity of the Biogas Plant

The required capacity of the biogas plant has been estimated after taking into account: (i) the reduction in biogas output in winter months[23], (ii) the tradeoff between capital costs of a larger biogas unit vis-à-vis the additional costs of diesel used in months when gas is not sufficient to meet the requirements of the critical months, and (iii) the availability of cow dung and ownership of cattle for different farm sizes.

A sample calculation for Ghazipur can be made as follows: a biogas plant of 2 m^3/day capacity produces enough biogas to pump the water requirement of a 1 ha farm except in October and November. The biogas is sufficient for the pumping requirement during the winter months of December, January, and February, even though there is a reduction of 39 percent in biogas output during these months. During October and November, when there is a deficit of biogas, the dual- fuel engine can use more than 30 percent diesel to meet the pumping requirement. The additional diesel needed is 13 liters in October and 42 liters in November. The cost of this additional quantity of diesel is estimated at Rs 248/yr. In contrast, if a 3 m^3/day biogas plant is used, it will provide enough biogas for all months except November. The additional diesel required under this case is 20 liters, costing Rs 90/yr. Similar calculations have been made for other crop rotations in Ghazipur taking into account various combinations of sizes of biogas plants. The size of the biogas plant to be installed is determined after comparing the (annualized) additional capital costs of larger biogas plants with savings in diesel costs.

Windmill Capacity

As stated earlier, an objective of this study is to assess the option of wind energy for irrigation and not to assess any particular windmill. It is therefore assumed that a "similar" windmill of a larger or smaller size can be constructed to suit the wind conditions and water requirement of a given place under consideration. It is also assumed that efficiencies remain unchanged in the range of sizes involved.

The critical month is identified the same way as in the case of solar systems: the month in which the ratio between water requirement and its availability is the highest (Table 6.2). As in the case of solar systems, care was taken to optimize the installed capacity in such a manner that water available during the year is satisfactorily used. Where rice is one of the crops, the area under rice cultivation is reduced, so that the critical month is determined by non-rice requirements. Using the estimated requirement and water output, the sizes of three types of windmills were determined as follows: (a) "appropriate technology" design, Ghazipur-type windmills, 5 m rotor diameter, with average overall monthly efficiency assumed at 10 percent; (b) hybrid technology design, NAL-MP-1, 7.5 m rotor diameter, with average overall efficiency of 5 percent (measured); and (c) conventional technology design (such as Comet, Southern Cross, etc.), with assumed efficiency of 10 percent. These efficiencies are assumed to be constant, regardless of wind regimes.

Capital and Operating Costs

Capital and Operating Costs of Photovoltaic System

The current capital costs of a photovoltaic system will vary across countries depending on the number of units purchased and the location. For

Table 6.7
Estimated Capital Costs for Photovoltaic and Solar-Thermal Systems (5 m head) in Ghazipur (in 1983 $/ha)

System and crop rotation	Current prices		Projected prices	
	at current efficiency	at projected efficiency	at current efficiency	at projected efficiency
Photovoltaic system[1]				
I wheat + rice	4,820	3,560	1,490	1,100
II wheat + bajra	4,820	3,560	1,490	1,100
III sugar-cane	13,870	10,250	4,280	3,160
Solar thermal system[2]				
I wheat + rice	5,370	2,690	-	-
II wheat + bajra	5,370	2,690	-	-
III sugar-cane	15,440	7,720	-	-

Note: Estimated capital costs depend on price levels and efficiency levels (current vs. projected), which are explained below for both systems. Costs of solar-thermal system only appear at current prices because these are not expected to change considerably in the future.

[1] For photovoltaic system:
—current price is $9.75/W_p$ for the arrays and $2/W_p$ for the subsystem (motor, pumps, and pipework).
—projected price is $2.50/W_p$ for the arrays and $2/W_p$ for the subsystem.
—current efficiency is 3.4 percent of overall system instantaneous value at 1000 W/m^2.
—expected efficiency is 4.6 percent of overall system instantaneous value at 1000 W/m^2.

[2] For solar-thermal system:
—current price is $200/m^2 for the thermal collectors and $86/m^2 for the subsystem.
—current efficiency is 0.5 percent daily average overall system in a 6 kWh/m^2.
—expected efficiency is 1.0 percent daily average overall system in a 6 kWh/m^2.

Source: Table 6.5 and notes.

Table 6.8
Present Values of Costs of Photovoltaic and Solar-Thermal Systems in Ghazipur (in 1983 $)

System and crop rotation		Current prices				Projected prices			
		current efficiency		expected efficiency		current efficiency		expected efficiency	
		farm size		farm size		farm size		farm size	
		0.4 ha	1.0 ha	0.4 ha	1.0 ha	0.4 ha	1.0 ha	0.4 ha	1.0 ha
Photovoltaic system[1]									
I	wheat+rice	2,110	5,000	1,610	3,740	920	2,030	730	1,550
II	wheat+bajra	2,110	5,000	1,610	3,740	920	2,030	730	1,550
III	sugar-cane	5,730	14,050	4,280	10,430	2,310	5,490	1,750	4,100
Solar thermal system[2]									
I	wheat+rice	2,550	5,870	1,450	3,030	-	-	-	-
II	wheat+bajra	2,550	5,870	1,450	3,030	-	-	-	-
III	sugar-cane	6,690	16,230	3,520	8,280	-	-	-	-

Note: The estimated capital costs on which the above present values are partly based depend on price levels and efficiency levels (current vs. projected), which are explained below for both systems. Figures for the solar-thermal system only appear at current prices because capital costs are not expected to change considerably in the future.

[1] For photovoltaic system:
—current price is $9.75/W_p$ for the arrays and $2/W_p$ for the subsystem (motor, pumps, and pipework).
—projected price is $2.50/W_p$ for the arrays and $2/W_p$ for the subsystem.
—current efficiency is 3.4 percent of overall system instantaneous value at 1000 W/m^2.
—expected efficiency is 4.6 percent of overall system instantaneous value at 1000 W/m^2.

[2] For solar-thermal system:
—current price is $200/m^2$ for the thermal collectors and $86/m^2$ for the subsystem.
—current efficiency is 0.5 percent daily average overall system in a 6 kWh/m^2.
—expected efficiency is 1.0 percent daily average overall system in a 6 kWh/m^2.

Source: Tables 6.5, 6.7, and text.

photovoltaic arrays, the average cost in 1984 was around $9/W_p$ (delivered cost). It is estimated that the cost of photovoltaic subsystems (motor, pump, and pipework) is about $2/W_p$.[24] There are a number of projections and expectations about prices of photovoltaic cells. Prices have not fallen significantly in the last five years, but a technological breakthrough could make it happen.[25] To take into account possible cost reductions, a projected price of $2.5/W_p$ for the year 1990 is assumed. Since costs of photovoltaic subsystems are not likely to decline, the combined future costs of photovoltaic systems are assumed to be $4.5/W_p$.

Capital costs will be lower by a factor of 0.74 if the technological advances raise the peak instantaneous system efficiency to 4.6 percent from the present 3.4 percent (with reference to gross cell area). Since cell arrays do not involve moving parts, there are no repair and maintenance and fuel costs. However, field experience has shown that maintenance of motor and pump as well as replacement of brushes, and so on may require an estimated expenditure of around $30/yr.[26] The estimated capital cost and sum of present values of capital and operating costs under various assumptions of prices and efficiencies are given in Tables 6.7 and 6.8 respectively.

Capital and Operating Costs of Solar-Thermal Systems

In the international market, flat plate collectors (single glazed, matt-black, nonselective absorber) are reported to cost around $110/m^2$ (1980) and the whole solar-thermal system should cost $170/m^2$ according to Halcrow et al. (1983). It is reported that a commercial firm in India is selling collectors in the price range of $170–$250/m^2$ (depending on the glazing and selective coating). Therefore for this study an average figure of $200/m^2$ for the collector is used. Assuming that the cost of the collector is 70 percent of the cost of the system, the latter will cost $286/m^2$. Using this, the capital costs of solar-thermal systems are estimated as varying between $5000 and $15,000 for a 1-ha farm in Ghazipur. The repair and maintenance costs of a solar thermal system will be equal to or higher than those of a diesel engine. Given the lack of any field data on this, a notional figure of $30/yr is used, although this cost will vary significantly from one country to another. The present values of costs for different crop rotations are given in Table 6.8.

Capital and Operating Costs of Electric and Diesel Pumpsets

The relevant cost items are the capital costs of an electric motor or a diesel engine (or dual-fuel engine) plus the cost of the actual pump and the costs of fuels (electricity/diesel) as well as the costs of operation and maintenance. However, the fact that almost all mechanized irrigation in India is being powered by electricity and diesel makes it necessary to distinguish between market prices and shadow prices of both capital and operating costs. Market prices do not reflect real resource costs of supplying these fuels to rural areas. In most states, including the Ghazipur region, farmers are not charged for these costs, implying that electricity and diesel are subsidized. In a realistic evaluation of alternative technologies, it is imperative that all the explicit and implicit costs incurred by society are taken into account. It is difficult to make accurate estimates of costs of infrastructure that can be attributed to agricultural uses, since these services are available to the society at large. Even if inaccurate, however, it is important to provide an idea of such costs if a policy choice between centralized or decentralized energy sources is being considered, where the latter does not require the same level of infrastructure facilities.

For the alternative where electricity is obtained from the regional grid, it is necessary to estimate the capital costs of generation and transmission attributed to the agricultural load; the cost of electrifying a village, which includes costs of distribution lines from the point where the tapping is done on the main high-voltage transmission lines; the costs of transformers; and the cost of connection from the nearest pole to the well/tubewell. According to our estimates, the maximum operating hours for a 2-hp (1.5 kW) electric pumpset to provide irrigation to a 1-ha farm in Ghazipur are 800 hrs/yr. This indicates a rather low load factor of 9 percent. The monthly distribution of these operating hours shows the seasonality of demand for power, which has to be taken into account in estimating the capacity costs of electricity used in the agricultural sector. For example, 0.5 million pumpsets to be energized during 1980–1985 in Uttar Pradesh and Bihar will require 2000 MW of additional capacity if they are used simultaneously. Because of variations in cropping patterns and rainfall and farmers' preferences, however, the actual generation capacity required will be lower. Assuming this is 1000 MW, the investments required will be Rs 10,000 million or Rs 20,000 ($2000) per pumpset at current cost levels ($1000/kW) (other uses of electricity are discussed below). To this, one must add about Rs 10,000 per pumpset to account for the costs of extra high voltage transmission lines, low-voltage distribution lines, transformers, and so forth. However, this entire cost cannot be solely attributed to irrigation pumping because the created power capacity will be usable by other consumers when it is not being used for pumping. One tends to think that in an economy where acute power shortages are reported, a substantial part of this off-peak (seasonal as well as during the day) power will be usable. However, this need not always be the case since power demands from agro-processing, agro-based industries, manufacturing units, and other consumers may not be forthcoming in the nonagricultural season. An accurate analysis of this issue requires information on load profiles, which is not readily available. Some of these off-peak demands, especially in the nonagricultural season, may be induced by charging lower tariffs during this period. Only half of the capital costs of generation and transmission (i.e., only Rs 15,000 per 4 kW pumpset) is attributed to irrigation pumping in this study, assuming that part of the capacity created for meeting the agricultural load can be used for other purposes.

To this figure, one may add the estimated capital costs of electrifying a village to provide power for pumping (among other uses) as well as the costs of connection to the well/tubewell. This is about Rs 7440 per pumpset if the entire cost of rural electrification is attributed to pumpset energization. However, since electricity can also be used by rural households (although these will be a small and relatively rich minority), agro-industries, and commercial establishments, a part of this cost should be attributed to other uses.[27] An average social cost of distribution and connection of Rs 5000 per pumpset is assumed in this study due to the lack of reliable data on this. Thus a total cost of generation, distribution, transmission, and connection of Rs 20,000 per 4 kW pumpset is estimated for the reference case. The implications of using a higher figure of Rs 37,000 per pumpset are discussed later in this section.

This estimate may be compared with others available in the literature. A recent paper[28] shows that the average capital investment by the government during 1985–1990 for energizing each of the additional 3 million pumpsets is approximately Rs 37,000. It is estimated that long-term marginal costs of providing power for irrigation pumping in Orissa (among other uses) are Rs 281/kW/month (Rs 13,500/yr for a 4 kW pumpset) or Rs 4.215/kWh at 9 percent load factor (800 hrs/yr). With a higher load factor (20 percent or 1750 hr/yr) the cost of electricity will be Rs 1.926/kWh, including capital cost and annual costs of operation, maintenance, and administration (OMA). Fuel costs have been estimated at Rs 0.287/kWh using a shadow price of Rs 170/t of coal and a conversion factor of 0.79 kg of coal to generate 1.0 kWh of electricity.[29] It is estimated that of the total capacity costs of service (Rs 281/kW/month), 25 percent are accounted for by annual OMA costs. For a load factor corresponding to 800 hrs/yr, this indicates an OMA cost of Rs 1.02/kWh. A lower figure of Rs 0.50/kWh is used here, however, since the load factor in Ghazipur is expected to be higher than in Orissa. This, added to the estimated fuel cost of Rs 0.287/kWh, gives an estimated operating cost of Rs 0.787/kWh for electricity used in pumping water. This cost, plus the capital cost of generation, transmission, distribution, and connection, gives us the "real" costs of providing electricity to rural areas for pumping. Obviously these costs are much higher than the tariffs charged to farmers. For example, it has been reported that, in Orissa, estimated tariffs were only 4 percent of the long-term marginal cost of providing electricity for irrigation.[30]

The cost of infrastructure for distributing diesel for pumping is very difficult to assess, especially because it comprises an uncertain share in the capital costs of refineries and/or port facilities. The cost-share in port facilities cannot be easily estimated; it is likely to be very small, and may be ignored since it cannot possibly have any significant effect on final cost comparisons of energy options for pumping. It is important, however, to take into account (i) the capital cost incurred in refining crude oil to obtain diesel oil, and (ii) the capital and operating costs of transporting of additional diesel by the railways and truck-tankers. The capital costs of refining capacity are incurred only if it is found to be economical to import crude oil and refine it in the country to obtain diesel together with other oil products. For India's conditions, however, it is more economical to import diesel (alone or with kerosene) than to import crude oil.[31] Hence, the shadow price of diesel used in this study is estimated as the c.i.f import price of diesel increased by the 25 percent premium on foreign exchange to reflect its overall scarcity.

Table 6.9
Capital Costs of Conventional Systems (in 1983 $)

	Electric pumpset (with electricity from regional grid and extension of line to village)		Diesel pumpset (or dual-fuel engine)	
Cost items / Size of motor	1 hp	2 hp	2 hp	3 hp
Engine/motor[1] (including pump)	273	301	400	486
Pump house (for electric pumpset)	100	100	-	-
Connection (to nearest pole)	500	500	-	-
Generation and transmission of electricity	400	800	-	-
Infrastructure for diesel transport	-	-	50	50
Total[2]	1,273	1,701	450	536

[1] Data from regional distributors of Kirloskar Oil Engine and Kirloskar Brothers Limited, New Delhi, for engines and motors.

[2] For details on cost items other than cost of engine/motor, see SDCA (1985) and Bhatia (1980a). Also see text.

Source: See notes.

Table 6.10
Annual Operating Costs of Conventional Systems (in 1983 $)

	Grid electricity		Diesel oil	
Size of motor/engine:	1 hp	2 hp	2 hp	2 hp
Size of farm:	0.4 ha	1.0 ha	0.4 ha	1.0 ha
Energy costs				
Electricity[1]	16	40	-	-
Diesel oil[2]	-	-	54	137
Lubricating oil[3]	-	-	5	12
Repair and maintenance costs				
Electric motor	30	30	-	-
Diesel engine	-	-	45	45
Labour charges for operation	-	-	30	30
Total annual operating costs	46	70	134	224

[1] Electricity: $0.0787 per kWh (shadow price as discussed in the text).
[2] Diesel: $0.45/l (assuming import cost of $0.3/l, 25 percent premium on foreign exchange, and $0.075/l for local transport.
[3] Lubricating oil is used in diesel engines and is therefore included in operating costs.

Source: Author's estimates; see text for explanations.

Table 6.11
Present Value of Costs for Irrigation Pumping Using Electricity from the Regional Grid and Diesel Oil for Ghazipur (in 1983 $)

Energy option and crop rotation	Farm size: 0.4 ha	Farm size: 1.0 ha
Electricity from grid		
I wheat and rice	1,555	2,129
II wheat and bajra	1,502	1,570
III sugar-cane	1,601	2,245
Diesel oil		
I wheat and rice	1,277	1,827
II wheat and bajra	1,078	1,333
III sugar-cane	1,450	2,260

Note: Present values have been estimated assuming a SRD of 10 percent and an equipment life of 10 years.

Source: Elaborated from data in Tables 6.9 and 6.10.

Table 6.12
Capital and Operating Costs of a Dual-Fuel Engine and Pump Using Diesel Oil with Biogas or Producer Gas in Ghazipur (in 1983 $)

	Dual-fuel engine and pump	
Cost items	Diesel and biogas 2 hp (1 ha)[1]	Diesel and producer gas 3 hp (1 ha)
Capital costs		
Dual-fuel engine and pump	400	486
Infrastructure for diesel transportation	50	50
Biogas plant (by size)[2]		
2 m^3/day	400	-
4 m^3/day	-	-
Gasifier 5 hp	-	450
Sum of capital costs	850	986
Operating costs		
Fuel costs		
diesel[3]	65.8	41
lube oil	5.8	4
cow dung[4]	-	-
fuelwood[5]	-	30
Repair and maintenance		
dual-fuel engine and pump	45	45
biogas plant	10	-
gasifier (15 per cent of capital cost)	-	68
Labour charges for operation of dual-fuel engine	30	30
Sum of operating costs	156.6	218

[1] Area that can be irrigated.
[2] See text about determination of size.
[3] The cost of diesel is estimated as follows: For a dual-fuel engine using biogas there are two components: (0.135 liters/hp-hr) × (680 hp-hrs/year) × ($0.45/l) = $41.0 plus $24.8, which is the cost of 55 liters of additional diesel required in October and November due to the unavailability of gas (see text p. 183). For a dual-fuel engine using producer gas from a gasifier, the diesel cost is only $41.0 per year since the gasifier (5 hp) is of adequate size.
[4] Cost of cow dung is not included since the slurry that comes out of the biogas plant is at least as good a manure as cow dung (if not better). Hence, there is no resource cost in using cow dung as input.
[5] Assuming 3.5 kg of fuelwood per liter of diesel oil saved. If gas is not used, a dual-fuel engine will use 0.45 l/hp-hr. When gas is used, diesel saved is 0.315 l/hp-hr. This gives a consumption of 3.5 × 0.305 = 1.1 kg of fuelwood per hp-hr. For an annual pumping of 680 hp-hrs, the total fuelwood required is 0.75 tons costing $30.

Source: Author's field work; see text for assumptions.

Capital and Operating Costs of Dual-Fuel Engine and Biogas Plant

The estimated costs[32] for selected cases are shown in Table 6.12. Similar costs have also been calculated for other crop combinations. The cost of cow dung is not included since the slurry which comes out of the biogas plant is at least as good (if not better) a fertilizer as cow dung. Hence, there is no resource cost in using cow dung. Present values of costs for the system under various assumptions are in Table 6.13

Capital and Operating Costs of Dual-Fuel Engine and Small-Scale Gasifier

These costs are shown in Table 6.12. It is assumed that a 5 hp gasifier costs $450 (at 1983 prices) and with derating (to 70–80 percent), a 3-hp dual-fuel engine is able to provide enough pumping capacity to irrigate more than 1.0 ha. The life of a gasifier is estimated at 10 years and the annual repair and maintenance costs at 15 percent of capital costs. The opportunity cost of fuelwood is most difficult to estimate. It is reported that market prices of fuelwood in semi-urban areas of Uttar Pradesh ranged between $40/t and $45/t in 1983/84. The use of wood for paper mills and construction poles can fetch a higher price. An additional demand for fuelwood by gasifiers may result in overall shortages and may lead to deforestation and high environmental costs. Besides, true opportunity costs of land and unskilled labor are expected to vary from region to region and season to season. A detailed analysis of these factors is beyond the scope of this study. Hence, a

Table 6.13
Present Values of Costs of Using Diesel Oil and Biogas for Irrigation Pumping in Ghazipur (in 1983 $)

		Farm size	
Crop rotation		0.4 ha	1.0 ha
I	wheat and rice	1,480	1,810
II	wheat and bajra	1,420	1,500
III	sugar-cane	1,530	1,920

Note: Assuming an SRD of 10 percent and an equipment life of 10 years.
Source: SDCA (1985).

Table 6.14
Present Values of Costs of Using Diesel Oil and Producer Gas for Irrigation Pumping in Ghazipur (in 1983 $)

	Farm size	
Crop rotation	0.4 ha	1.0 ha
I wheat and rice	2,040	2,313
II wheat and bajra	1,939	2,067
III sugar-cane	2,124	2,528

Note: Assuming an SRD of 10 percent and an equipment life of 10 years.
Source: See Table 6.12 and text for assumptions.

national price of $40/t of fuelwood is used to reflect its social cost.[33] It is further assumed that a dual-fuel engine will use 0.135 l/hp-hr of diesel and 1.1 kg of fuelwood (assuming 3.5 kg of wood replaces 1 liter of diesel and 0.450 1/hp-hr of diesel is consumed if gas is not used). Table 6.14 shows the present values of the costs of irrigation using producer gas and diesel. For a 1-ha farm using crop rotation I, the present value of costs is $2313 when the economic cost of fuelwood is $40/t. The present value decreases to $2170 if the economic cost of fuelwood production is $10/t.

The additional equipment required to transport diesel (rail- or truck-tankers) should be explicitly taken into account. A detailed calculation of the various components of costs is beyond the scope of this study and, hence, only a notional figure is used. For an additional 2.0 million diesel pumpsets during the Sixth Plan, requiring about 0.5 m.t of diesel, the additional capital costs to be incurred may be about Rs 1000 million, or Rs 500 per pumpset.[34] These are added to the costs of diesel pumpsets as infrastructural capital costs. The operating expenses of these tankers (including the cost of electricity and diesel required) are added to the estimated price of diesel oil. Based on the 1981/82 import price of high-speed diesel and light diesel oil, a c.i.f import price of Rs 3/l is estimated. The 25 percent premium on foreign exchange raises it to Rs 3.75/l. This is increased by a notional figure of Rs 0.75/l, which may have to be incurred in transporting diesel to distant places (over and above the capital costs of tankers, etc.). Hence, a shadow price of Rs 4.5/l is used in this study. The capital costs of these alternatives are given in Table 6.9, the operating costs are in Table 6.10, and the present values of costs for electric and diesel pumpsets are in Table 6.11.

Capital and Operating Costs of Windmills

Estimated capital costs of the three types of windmills of equivalent efficiency and designed to provide irrigation to a 1-ha farm are given in Table 6.15. These are a conventional and reliable windmill, costing $200/m^2; an intermediate hybrid technology costing around $90/m^2; and the least expensive, Ghazipur multivane-type windmill, costing $75/m^2. In terms of maintenance requirements, the experience with NAL's WP-2, another conventional multivane windmill, has shown that rotor and top housing require practically no attention. But the pump and the plunger rod have to be looked after. No more than 0.1 percent of capital cost may be assumed for the cost of maintenance of NAL's MP-1 windmills. The Ghazipur-type windmill requires significant attention. Sinha (1982) reports that the maintenance cost of this windmill is about Rs 500/yr. A variety of items including crank, cross-head, pins and bushes, wooden pump, rod guides, piston valves, and sleeve in the tower pipe have to be replaced at least once a year. Even the main shaft bearing may have to be replaced once every five years. Exposed bearings and some other parts have to be lubricated every 15 days. Assuming that all the routine lubrication is carried out by the farmer himself, a mechanic may have to visit once every three months for a replacement of pump valves. It appears, therefore, that the stated maintenance cost of Rs 500 may be a bare minimum and could, in fact, be higher. Thus, the annual maintenance cost is 4 percent of its capital cost. Also, in the sixth and eleventh years of use of the windmill, main shaft bearings have to be replaced, which may cost an additional Rs 1000 in those years.

In the MP-1, windmill canvas sails (if used) may have to be replaced every two years, costing Rs 800 including nonmetallic ropes for the rotor. However, the experience with polypropylene sails suggests that these easily last for two years and cost only about Rs 500. Therefore, it is assumed that the replacement cost of sails is Rs 500 every alternate year, which amounts to 2.5 percent of the capital cost. The remaining components in the MP-1 windmill are properly sealed, requiring periodic checks every three months for oil replacement. For this, an expenditure of Rs 100 per year or 0.2 percent of the capital cost is assumed. Apart from this, there should be no need for periodic replacement of windmill parts. However the pump, which costs Rs 1000 (1983 prices), needs to be replaced once every 10 years or so.

Table 6.15 gives the estimated present values of economic costs over a period of 10 years for 1.0 ha of cultivation. In all cases, the Ghazipur windmill has the lowest costs. However, since it requires more frequent maintenance, its supply of energy may be lower and its field performance may deteriorate progressively with use.

Table 6.15
Estimated Capital Costs and Present Values of Costs per ha at Ghazipur for Three Types of Windmills (in 1983 $)

Cost items	Crop rotation	Windmills[1]		
		A	B	C
Capital costs	I	3,840	5,660	10,240
	II	4,200	6,190	11,200
	III	8,400	12,390	22,400
Present value of total costs (for ten years' operation at 10 per cent SRD)	I	4,960	6,060	10,300
	II	5,420	6,620	11,270
	III	10,840	13,260	22,530

[1]The following windmills are considered:

A: "Appropriate-technology" design: Ghazipur/Allahabad 5-m rotor diameter, costing $1200 ($75/m^2 approx.). Average overall monthly efficiency (assumed) is 10 percent.

B: Hybrid technology design: NAL MP-1, 7.5-m rotor diameter, windmill costing $2000 ($45/m^2 approximately). Average overall efficiency (measured) is 5 percent.

C: Conventional technology design, such as Comet, Southern Cross, and so on, costing $200/m^2. Assumed efficiency is 10 percent, as for type A.

Source: SDCA (1985).

A Comparison of Present Values of Economic Costs

A comparison of the sum of present values of capital and operating costs of alternative systems at Ghazipur for two farm sizes (1.0 ha and 0.4 ha) is presented in Table 6.16.

For the 1-ha farm, the use of diesel and biogas in a dual-fuel engine is the most economic option[35] for crop rotations I and III, except the solar-photovoltaic system under assumption D (projected system cost $4.5/W$_p$ and projected 4.6 percent instantaneous efficiency). For this 1-ha farm under crop rotation II, the diesel-fuel engine ranks second, next to the diesel pumpset.

The use of electricity from the regional grid is consistently more expensive from the viewpoint of society when the real resource costs of supplying electricity to rural areas are taken into account. Although the differences in present values of costs between the biogas-cum-diesel option and conventional sources range from 10 to 25 percent, the present analysis shows the need for a thorough investigation of the socioeconomic and

Table 6.16
Present Values of Sum of Capital and Operating Costs of Alternative Technologies for Irrigation Pumping in Ghazipur (in 1983 $)

		Farm size: 1 ha Head: 5 m			Farm size: 0.4 ha Head: 5 m		
		Crop rotation			Crop rotation		
Technologies		I	II	III	I	II	III
Electricity		2,130	1,570	2,250	1,550	1,500	1,600
Diesel oil		1,830	1,330	2,260	1,280	1,080	1,450
Biogas + diesel		1,810	1,500	1,920	1,480	1,420	1,530
Producer gas + diesel		2,310	2,070	2,530	2,040	1,940	2,120
Solar photovoltaics	(A)	5,000	5,000	14,050	2,110	2,110	5,730
	(B)	3,740	3,740	10,430	1,610	1,610	4,280
	(C)	2,030	2,030	5,490	920	920	2,310
	(D)	1,550	1,550	4,100	730	930	1,750
Solar thermal	(E)	5,820	5,380	16,230	2,550	2,360	6,690
	(F)	3,030	2,790	8,230	1,450	1,350	3,520
Windmills	(G)	4,960	5,420	10,840	1,980	2,170	4,340
	(H)	6,060	6,620	13,260	2,420	2,650	5,300

Note: The figures in this table have been rounded up to the nearest ten for easy comparison. Rows (A) through (H) include different assumptions as noted below.
(A): Current costs (\$11.75/W_p); current efficiencies (3.4 percent overall instantaneous).
(B): Current costs (\$11.75/W_p); future efficiencies (4.6 percent overall instantaneous).
(C): Future costs (\$4.5/W_p); current efficiencies (3.4 percent overall instantaneous).
(D): Future costs (\$4.5/W_p); future efficiencies (4.6 percent overall instantaneous).
(E): \$286/m^2 for the whole system and 0.5 percent daily average efficiency.
(F): \$286/m^2 for the whole system and 1.0 percent daily average efficiency.
(G): Appropriate technology (Ghazipur-Allahabad multivane type with a piston pump).
(H): Hybrid technology (NAL sail type with a centrifugal pump).

Source: Tables 6.8, 6.11, 6.13, 6.14.

management aspects of using biogas for irrigation pumping. In particular, pricing and tax/subsidies on energy sources must be studied carefully. Given the subsidized rates of electricity and no charges for electricity connections, farmers prefer grid-electricity to diesel pumpsets and dual-fuel systems using biogas and diesel. They might not prefer electric pumpsets if they were charged the real costs of providing grid electricity. If this were the case, investments in a biogas plant along with a dual-fuel engine might be financially feasible. However, if subsidies on electricity cannot be removed, there is a need to give an equivalent subsidy to biogas plants used for irrigation. Some of these aspects are discussed in the next section.

The costs of producer gas (from a gasifier) with diesel oil are found to be higher than those for grid electricity and diesel/biogas options, primarily because of high costs of repair and maintenance and of fuelwood.[36] However, apart from the high costs, the technical parameters of using wood-based (and even charcoal-based) gasifiers in India have yet to be firmed up. There is a need to sponsor a few field studies to document various techno-economic aspects of using gasifiers for pumping.

The photovoltaic system, under the lowest-cost and highest-efficiency assumptions (case D), is much cheaper than the average of the first four alternatives in respect of crop rotations I and II. It is much more expensive for crop rotation III (sugar cane) because the capacity for meeting water requirements in the critical month (May) is very high. At current prices, the present values of costs of photovoltaic systems are 2–2.5 times higher than those of other alternatives (for crop rotations I and II). In the intermediate situation of future costs and current efficiencies (case C), the solar photovoltaic option is cheaper than grid electricity for crop rotation I. The best case for solar-thermal systems is more expensive than the best of the photovoltaic systems for crop rotations I and II.

Windmills are 2–3 times more expensive than conventional systems. The Ghazipur windmills are almost as expensive as the photovoltaic system at current costs. The hybrid technology windmill is much more expensive than the photovoltaic system. In fact, this shows that areas like Ghazipur in northern India are not suitable for windmills because of their inadequate wind conditions.

For the 0.4-ha farm, under crop rotations I and II, the photovoltaic system is cost competitive with conventional energy options at the assumed future prices and current and future efficiencies (cases C and D; see Table 6.16). The photovoltaic option is economic even under current costs and better efficiency (case B) if the estimated costs of grid electricity are higher by about Rs 600 for the same crop rotation. For this 0.4-ha farm, the photovoltaic system is economic even under current costs and current efficiency for regions/farms where the cost of extending grid electricity is higher (by Rs 5,600 per pumpset) than that estimated in this study.

Solar-thermal with higher efficiency (case F) also becomes attractive for crop rotations I and II in Ghazipur. Biogas with diesel remains competitive with conventional energy options but producer gas with diesel is more expensive in the 0.4-ha case. Even for crop rotation III, which requires relatively more pumping, the savings in diesel oil do not outweigh the capital costs of biogas plant/gasifier or extension of grid electricity.[37]

The above discussion is based on the assumption that the average costs of providing grid electricity are on the order of $2000 per pumpset (of 4 kW). The results regarding renewables, especially for photovoltaics, will change if an average figure of $3700 per pumpset is taken as estimated by the Rural Electrification Corporation for the Seventh Plan (1986—1990). Such costs are incurred under a variety of situations relating to distance from the nearest distribution network and connected load. For example, assuming a load of 20 kW, the costs are about $18,500 if a low-tension line is extended beyond a distance of 10 km. Since there are villages in the northern plains that are more than 10 km away from the existing electricity distribution system, it is interesting to review the results in the context of a higher figure for grid electricity costs. According to our calculations, if a figure of $3700 per pumpset is assumed, the following conclusions are reached:

i) under case C (future costs, current efficiencies), the photovoltaic option is more economic than grid electricity for crop rotations I and II for a 1-ha farm;
ii) for a smaller farm (0.4 ha), the use of the photovoltaic system is more economical than the electric pumpset under extension of grid electricity, in all three crop rotations, even under assumption A (i.e., current costs and current efficiencies); and
iii) the use of gasifier and dual-fuel engine is also economic compared with grid electricity, for a farm size of 0.4 or 1.0 ha.

Thus, to recapitulate, the use of biogas and diesel for pumping is found to be economic under a variety of conditions. Photovoltaic systems are economic for small farms even at the current high capital costs, when compared with the alternative of extending low-tension lines beyond 10 km. The use of gasifiers is also found economic where costs of grid electricity are high or diesel costs are high due to lower depth of water table. However, none of these alternatives will be adopted and used by the farmers under existing conditions of subsidies on use of electric motors and low diesel prices. The diffusion of these technologies requires different policies, which are discussed in the third section, below.

Significance of Factors Other than Economic Costs

Various sources/technologies differ from each other in terms of technological, social, environmental, and economic aspects.[38] Apart from economic costs, those other aspects must be considered in any realistic analysis of energy alternatives for irrigation pumping.

Among the technological aspects, levels of field experience and users' familiarity with technology are rather low in the case of renewables such as windmills, solar-thermal, and photovoltaic systems. Biogas technology is widely known but its use has been restricted mainly to cooking and lighting. The use of biogas and producer gas for running engines is not common, although dual-fuel engines (which are diesel engines with minor modifications) are likely to be accepted without much difficulty if gas supply is reliable and financial conditions are favorable. The social acceptance of biogas or any technology thus depends on whether or not it is considered dependable by users.

The reliability of equipment depends crucially on the complexity of the device and the availability of repair and maintenance facilities. It also depends on the quality of manufacturing and testing of machines as well as availability and quality of spare parts and skilled manpower. Standardization of equipment and spare parts has to be considered an essential component of the strategy of diffusion of new technologies. In this context, solar-photovoltaics have an advantage over other devices because they are simple, do not require much repair and maintenance, and are modular in structure for easy repair and replacement. Some designs of windmills, which do not require periodic maintenance, also belong to this category. Solar-thermal devices, on the other hand, are known for their complexity of operation and maintenance and cannot easily use the existing infrastructure for repairs and maintenance. Solar-photovoltaics also score high in terms of the possibility of a technological breakthrough, which would significantly improve performance and reduce costs. Prices of solar-thermal devices and windmills are much less likely to change.

Another issue that may be important to take into account in comparative evaluations of technologies is the risk of depending on large-scale, centralized sources (power plants, refineries) vis-à-vis using decentralized sources such as biogas plants, gasifiers, windmills, and photovoltaic systems.[39] Experience has shown that a shutdown in a large power plant or a defect in a main transmission line can affect large areas that have no recourse to an alternative source of supply. Such uncertainty can seriously affect farmers' decisions regarding cropping pattern, use of fertilizers, and so on. An actual shortage at crucial stages of cultivation may reduce output significantly. This is not to claim that the use of renewable sources is less risky than the use of conventional sources or that production activities using renewable sources are more reliable than those using conventional ones. In fact, the existing infrastructure for repair and maintenance of renewable energy technologies may not be adequate and, hence, these may be less dependable than conventional technologies. These considerations, however, emphasize the need for analyzing the reliability of pumping equipment, energy producing units (power plant, refinery, gasifier, biogas plant), and distribution system (distribution lines for electricity, marketing channels for diesel oil, etc.).

An additional issue worthy of consideration in comparative evaluations of technologies is that the direct and indirect environmental effects of the renewable energy options discussed above are generally more favorable than those of competing conventional systems.

Income distribution aspects

In the above analysis, income distribution issues have been incorporated indirectly by considering small and marginal farmers. There are other aspects that need to be analyzed explicitly in the context of choice of energy alternatives. These are:

- the direct and indirect employment implications of alternative sources, such as biogas plants, photovoltaic systems, or windmills and resulting access to income by different categories of workers;
- the possibilities of depriving poor people of their supplies of dung or crop residues or fuelwood for cooking when these inputs become more valuable for the rich who do not want to part with them; and
- the impact of commercialization of fuelwood on the incomes of the poor.

These implications have to be analyzed and quantified, wherever possible, both at regional and macro levels.

POLICY AND ORGANIZATIONAL ISSUES

Need for Reviewing Rural Electrification Policies

The above analysis suggests that there is a need for reviewing the policies and programs of rural electrification in India. From the viewpoint of society, decisions regarding the extension of grid electricity to rural areas for irrigation pumping need not be economical in a variety of circumstances. The capital costs of powering a pumpset may be much higher than those shown in conventional analyses.[40] The gestation period of extending grid electricity, involving setting up of generation, transmission, and distribution facilities, may be 5–10 years depending on the type of generation facility. This problem becomes more serious in the context of overall shortages of power in the economy. The experience of rural electrification in India[41] shows that serious shortages have developed in the power sector on account of various financial, managerial, and political reasons. Investments in the modernization of power plants and transmission/distribution facilities have not been made, resulting in inadequate power supply to rural areas. The reaction of farmers to power shortages has been to oversize their electric motors and invest in back-up systems. As a result, a majority of large and medium farmers own both a diesel engine and an electric motor.

There is a need for a selective approach in rural electrification. Each project should be analyzed in terms of estimation and utilization of load, distance from the grid, and costs of alternative sources. In fact, the approach to providing electricity to rural areas should be changed. The emphasis should be on "rural energization" and not on "rural electrification" per se. In such an approach, grid extension will be only one of the alternatives, which include decentralized electricity generation and direct use of biogas, pyrolysis gas, wind power, and solar energy.

Role of Energy Pricing and Tax/Subsidy Policies

As discussed earlier, prices for energy sources such as electricity and diesel oil are administered by the government, and these involve different elements of subsidy. For example, in the case of grid electricity, there is a substantial element of subsidy where a farmer is not required to pay for the capital cost of connection and the electricity tariff is a small fraction (on average about 15–20 percent) of the real cost of supplying electricity. In contrast, diesel oil attracts excise duty even though its market price is lower than its shadow price (c.i.f price adjusted for premium on foreign exchange). A farmer investing in a biogas plant receives a subsidy (about 25 percent) on capital cost and is entitled to a bank loan at reduced rates. Although a few demonstration units have been installed free or at low cost, there have been no explicit subsidies for investments in photovoltaic systems and windmills. The use of gasifiers and ethanol for irrigation has not even been considered for government subsidies. Bhatia (1985) estimated that the available subsidies on biogas plants are not adequate to make an investment in a biogas plant (for irrigation) financially feasible. Similarly, other renewables should receive subsidies equivalent to those being offered to electricity.

In view of these considerations, it is necessary to assess:

i) the amount of subsidies being provided to various users of electricity in different regions;
ii) the level of costs that will be incurred in providing electricity to new areas; and
iii) the amount of subsidies that will make investment in renewables socially profitable and financially feasible.

The objectives of this assessment are to determine the optimum mix of energy alternatives from the viewpoint of society; ascertain policy changes that would be required in energy pricing and tax/subsidy schemes; and indicate organizational requirements to promote the diffusion of these technologies.

Organizational Aspects

The pursuit of the three objectives outlined above should, in general, be entrusted to either the planning ministry or an inter-ministerial committee.

However, the development and implementation of programs and related research concerning renewable energy technologies should be coordinated by a specialized body.

In India, renewable energy investments and related activities are coordinated by the Department of Non-Conventional Energy Sources. The above economic evaluation and policy analysis suggest that such an apex body should sponsor studies for field-testing and ex-post evaluation of various technologies in order to analyze the various aspects that are important in their development and diffusion. To the extent possible, however, these studies should use a uniform methodology that permits a comparative appraisal of the technology options for given end-uses.

SUMMARY OF CONCLUSIONS

The conclusions of this study can be summarized as follows:

1. Based on available data and assumptions, the use of electricity from the regional grid for irrigation pumping is consistently more expensive from the viewpoint of society than diesel or diesel-cum-biogas. This conclusion suggests that there is a need for a more selective approach in rural electrification, one that would thoroughly evaluate assumptions regarding the utilization of load, the distance from a grid, and the costs of alternative sources.

2. The use of biogas in a dual-fuel engine is found to be economic under a variety of assumptions. Since this option is not being used very widely at present, it is important to analyze financial, institutional, and organizational aspects in detail. In particular, policies of pricing and tax/subsidies on energy sources must be studied since low prices of electricity and/or diesel result in low cash savings from the use of renewable energy technologies, thus making the latter financially unprofitable. The use of gasifiers is found economic where costs of providing electricity/diesel are high and the economic costs of fuelwood production and use are low. The use of biomass resources for irrigation can have significant effects on employment and income distribution and these should be assessed carefully for each project.

3. The use of a solar-photovoltaic pumping system is economic for the small farm of 0.4 ha, even under current (1984) costs and current efficiencies for regions/farms where the capital costs of providing grid electricity are higher than Rs 25,000 per pumpset of 4 kW. Thus, solar-photovoltaic systems may be considered as a viable alternative for areas that are away from the grid and have low load factors of demand. Solar-thermal systems will become economic when significant improvements in the technical efficiency of utilization of solar energy can be made.

4. Windmils are 2–3 times more expensive than conventional systems

because wind regimes at Ghazipur are not suitable. This underlines the fact that the economics of windmills are site-specific and that for a number of locations wind conditions should be analyzed in detail before a program for the use of windmills is launched.

5. It is necessary to carry out detailed comparative economic evaluations of alternative technologies for given end-uses and locations under a variety of assumptions regarding cropping patterns, farm size, availability of groundwater, and seasonal variations in solar radiation or wind speeds.

6. Although some information on field testing of renewable energy devices is available, these data need to be strengthened by a proper evaluation of these systems. It may be necessary to provide for a built-in mechanism of simultaneous and ex-post evaluation as a part of each project on renewable technologies. In addition, it may be desirable to create an apex body that can coordinate the activities of various organizations connected with renewable energy projects, although this should be done in subordination to overall planning efforts at the national level.

NOTES

1. An outlay of Rs 1,860 crores ($1800 million) was provided during 1980–1985 to extend electricity to over 100,000 villages and to energize 2.5 million pumpsets. Besides, large-scale investments are required in electricity generation and transmission, which may total about Rs 10,000 crores ($10,000 million) for energizing 2.5 million pumpsets. The target for the Seventh Plan (1985–1990) is to energize 3 million additional pumpsets at an estimated cost of Rs 3000 crores ($3000 million). The corresponding investments in generation and transmission may be on the order of $15,000 million. Investments in rural electrification provide electricity for pumpsets but also for lighting and other uses, particularly to higher-income groups. For a discussion on socioeconomic aspects of rural electrification, see Fluitman (1983), Bhatia (1983c), Smith et al. (1983).

2. The array comprises a number of solar cells, arranged on a flat plate, and interconnected in series and parallel to provide the desired voltage and current. Typically, a single silicon cell provides an open circuit voltage of 0.6 V, which is independent of cell area, and the short circuit current is 25–30 mW/cm^2 at 100 mW/cm^2 insulation. At the operating temperature the open circuit voltage is about 0.55 V/cell. The power delivered by a cell depends on the cell area and solar radiation. In peak sunlight conditions represented by 100 mW/cm^2, the maximum power delivered by a cell (in terms of per unit cell area) is 12.7 mW/cm^2 at 0.48 V. The current generated remains near its peak value for load voltage up to 0.4 V/cell. The current and the power ratings drop proportionately at lower levels of radiation. Increase in operating temperature results in lowering of efficiency while at the same time reducing the operating voltage at maximum power. The figure of 12.7 mW/cm^2 also means a cell efficiency of 12.7 percent. However, when cells are packed together in an array with some nonactive space between adjacent cells, the efficiency of the array as a whole is lower: about 5.8 percent when cells are circular in shape and somewhat higher with square cells.

3. The average array conversion efficiency was about 10.2 percent but 12 percent is expected. The average efficiency of the MPPT was 92 percent. The average efficiency of the motor and pump had been 32 percent. The average overall efficiency of the system was about 2.8 percent, but 6 percent is expected.

4. A notable exception is a recent report on a solar thermal system at Bhavnagar (Kishore et al. 1983) that operated satisfactorily for 100 hours during a period of 20–25 days (19 May to mid-June 1983). The total period of operation was 200 hours during February– June 1983. Total daily water pumped was constant at 6500 liters, irrespective of changes in collector type or area. It is concluded that on clear days the pump can operate with a daily overall efficiency (solar to hydraulic power) of about 0.45 percent.

5. This system utilizes heat-pipe collectors that are also more efficient than flat plate collectors but are more expensive.

6. For a techno-economic description and evaluation of biogas units in developing countries, see Bhatia (1980a), Bhatia and Niamir (1979), Ghate (1980), Makhijani (1977), Reddy and Prasad (1979), Barnett et al. (1978), Bhatia (1977), French (1979), van Buren (1979), Moulik (1982), Stuckey (1983) and Lichtman (1984b).

7. Reddy and Prasad (1977) have also discussed the use of biogas directly through a Humphrey pump without going through the intermediary of a diesel/petrol engine. Efficiencies of about 20 percent from fuel to water output were attained and the pump seemed suitable for low head, high discharge applications. However this option has not been considered due to lack of field data.

8. This analysis took into account shadow prices or economic costs of alternative energy sources (diesel, electricity, and photovoltaics) including distribution costs.

9. This design has been introduced by Khadi and Village Industries Commission, Government of India. It is the most common in India.

10. See also Tewari (1978, 1981a, 1981b) on windmills in India.

11. The power of the wind is proportional to the cube of the wind speed. The formula is given by $Q = 0.79\ V^3 R^2 hr/H$, where Q is output of water in m^3/hr, V = wind speed in m/sec, R = radius of wind wheel in meters, hr = hours, and H = elevation head in meters.

12. The *Rabi* season is the winter agricultural season. Major crops are wheat and barley, which are sown in November/December and harvested in March/April.

13. The benefit/cost ratio is greater than unity only if the use of irrigation is economic. However, this may not be the case in some regions (e.g., high rainfall areas of eastern India) and for some crop rotations. Hence, after the alternative technologies are screened and ranked on the basis of the methodology outlined here, it will be necessary to estimate the overall benefit/cost ratio taking into account the costs of irrigation and other inputs.

14. For Ghazipur, three crop rotations considered are (i) rice and wheat (later variety, December planting), (ii) bajra and wheat (November planting), and (iii) sugar cane (planted, 12 months).

15. Under this methodology, one first determines the evapo-transpiration rate (ET_0), for a reference region (i.e., "the rate of evapo-transpiration from all extensive surfaces of 8-15 cm tall, green grass cover of uniform height, actively growing, completely shading the ground and not short of water"). This parameter is determined on the basis of climatic data such as temperature, humidity, wind, sunshine, radiation, and so forth. The ET_0 is determined for mean climatic data of 10 or 30 days and

expressed in terms of mm/day. The effect of crop characteristics on crop water requirement is given by: $ET_{crop} = K_c(ET_o)$, where ET_{crop} is the evapo-transpiration for the crop and K_c is the crop coefficient. The effect of local conditions and agricultural practices such as soil water availability, salinity, method of irrigation, size of farms, and so on are also considered. With the help of ET_0 values calculated from climatic information for Ghazipur, K_c values for the three crop rotations considered have been obtained from the literature and the knowledge of agricultural practices in the region (for details, see SDCA, 1985, Ch. 2). In considering the influence of rainfall on water requirement, the mean rainfall over a period of about 30 years is used in this analysis to represent normal conditions. (Although the results reported here are based on assumptions of normal rainfall, the analysis also considers situations that represent severe drought conditions, i.e., using the lowest recorded rainfall data; for details, see SDCA, 1985). A rainfall effectiveness factor (i.e., water available to the plant in its root zone) is also selected in order to estimate crop water requirement. Since precise data on this factor are not available, the following is used: So long as the monthly rainfall is under 100 mm, one may assume 100 percent effectiveness factor, but this will be less in cases where soil is dry and with poor vegetative cover. Due to the lack of more reliable primary data on rainfall effectiveness, a 70 percent effectiveness factor is assumed when the rainfall exceeds 100 mm. This assumption is based on an analysis of data published by the United States Department of Agriculture (FAO, 1977, p. 75). It is assumed that the contribution from groundwater (through capillary action) is negligible and water requirement for the purpose of leaching soil salinity is ignored. These factors are nevertheless secondary, since they do not bear significantly on the comparative assessment of various energy technologies.

16. See Dastane (undated) for details.

17. The implication of this is that groundwater provides only supplemental irrigation and the farmer, therefore, has to either obtain water from a surface water source for irrigating rice or cultivate it on a reduced field area. This, in fact, is known to be the practice in this region.

18. The total head is estimated as a sum of: (i) average static water level (below ground level), (ii) drawdown required to get estimated discharge, (iii) delivery head above ground level, and (iv) friction losses in pipe fittings, valves, and so on. The estimated total head for Ghazipur is 5 m.

19. For Ghazipur, for example, the irrigation water requirement for February (for crop rotations I and II) is estimated at 1390 m^3. This multiplied by the estimated 5 m head and again multiplied by 2.724×10^{-3} gives a figure of 18.93 kWh/month. The estimated gross cell area is 18.93 m^2 divided by 4.232, which equals 4.47 m^2. For details, see Halcrow et al. (1984).

20. It is estimated that under low discharge and low head conditions, a minimum pump efficiency of 50 percent is possible. Variation in head due to seasonal fluctuation of water availability should not reduce efficiency by more than 10 percent. In the case of belt-driven pumpsets, it may be necessary to take account of transmission efficiency, which may be 85 percent. In pumps driven by electric motors and diesel engines using direct flexible couplings, the transmission efficiency is close to 100 percent.

21. ARDC (1980) gives guidelines for selecting diesel engines and electric motors under varying discharge levels, total head, and pipe diameters. Based on

these guidelines, this study provides for extra bhp over the calculated bhp (of the electric motors and diesel engines) as follows: 50 percent increase for 2 bhp or less, 30 percent increase for units between 2 and 3 bhp, and 20 percent increase for units of greater capacity.

22. The smallest electric motor available in the market was 0.5 hp while the smallest diesel engine was 2 hp. This has been taken into account in the estimation of capital costs.

23. According to ICAR (1978), the output of gas in winter months was lower by 39 percent in the northern region, 31 percent in the western region, and 13 percent in the southern region. These estimates have been used to estimate the availability of gas in different months.

24. It is estimated that in the Indian Solar Photovoltaic Demonstration Programme the cost of motor and pump supplied were around Rs 5000 for a 300 W_p unit in 1982. Thus, an average cost of Rs 20 per W_p is reasonable. It is argued that there are no appreciable economies of scale in subsystem costs. See SDCA (1985).

25. A technological breakthrough could be the use of amorphous silicon cells. This could reduce cost of arrays significantly. The unit costs also depend upon the scale of manufacture. These may decline if demand picks up.

26. This is an average cost for one visit per month in India. This cost is much higher in regions where skilled labor is scarce. For example, in Fiji, 10–12 visits to the photovoltaic system could cost around $200.

27. It is beyond the scope of this study to discuss the issues relating to evaluation of rural electrification programs and patterns of utilization. See Smith (1980), Samantha and Sundaram (1983), Smith, Hayes, and Mehta (1983), Bhatia (1983c), Fluitman (1983), and Government of India (1983).

28. Prepared by the Central Electronics Limited on economics of rural electrification using photovoltaics, quoting from the Report of the Sub-Group on Rural Electrification for the Seventh Plan.

29. The cost of fuel, on these assumptions, is Rs 0.1343 (13.43 paise) per kWh. To measure the cost of fuel per kWh sold, account must be taken of station auxiliary consumption in the power plant as well as transformation, transmission, and distribution losses. At peak periods, the estimated fuel cost is Rs 0.287/kWh for low-voltage consumers.

30. Gellerson (1979) has estimated the ratios of tariffs to long-run marginal costs for various consumers in the northern, western, southern, and central regions. In the northern region, the agricultural tariff is 20 percent of the long-run marginal cost. The fuel costs for this case are 18.2 paise/kWh and the total costs 98.6 paise per kWh.

31. The economics of importing diesel oil vis-à-vis crude oil depends on the relative prices of these products, costs of refining, and demand pattern for products other than diesel. For a detailed study of these options, see, for example, Bhatia (1983b).

32. There are significant possibilities for reducing capital costs of biogas plants.

In fact, some improved versions of the KVIC design and units of Chinese design (Janata type) are already available. The Janata type plant is estimated to cost about 40 percent less. However, it is not clear whether the Janata design has the same level of gas output of KVIC design for equal capacity.

33. The estimation of economic cost of fuelwood in a region is a detailed study in itself. The most important consideration is the opportunity cost of land. Another is the possibility that the use of fuelwood and/or other biomass for gasifiers may deprive the poor of cooking fuel that they may presently be getting free.

34. Assuming that one truck-tanker makes one trip per week, transporting 500 t/yr, 1000 lorries are needed to transport 500,000 t/yr. This may require capital costs of about Rs 400 million compared with Rs 600 million for railway tankers moving the same quantity.

35. Of course this option assumes that a biogas plant will be installed primarily for providing gas to the dual-fuel engine. This involves transport of dung from the stable/house to the field and the problems vary from one farm to another. The management aspects of this alternative have to be studied through a number of field studies.

36. The cost of fuelwood may be low, however, if it is grown by the farmer on his own land, depending on opportunity costs of land and family labor.

37. In a recent study, Hurst (1984) noted that at current prices, large farmers (in northeastern Indian plains) should choose diesel engines over biomass technology or animal-powered pumps. He further noted that the problem with biomass technology in this region is that the groundwater is near the surface and the cost of diesel oil is not sufficiently large to compensate for the capital cost of biomass technology. However, at a 20–25 percent increase in the depth of the groundwater table, gasifiers fuelled with crop residues and fixed-dome biogas plants become more profitable than diesel-fuelled pumpsets.

38. For a detailed discussion of "appropriateness" of wind and other energy systems for irrigation water pumping in India, see Tewari (1981a, pp. 445–455).

39. Munasinghe (1979) provides a good discussion of this issue. There have been a number of studies on the question of reliability in power systems planning. Some of the approaches developed there could be used for evaluating the economics of decentralized systems.

40. The capital costs range from $2000 to $3700 for powering a 4 kW pumpset in northern India.

41. For a detailed discussion, see Bhatia (1983c).

7

Conclusions and Policy Implications

Armand Pereira and Ramesh Bhatia

This chapter begins with an overall summary of the main findings and their policy implications, followed by a brief discussion of their general implications for research and training.

SUMMARY OF FINDINGS

Macroeconomic Evaluation of Ethanol Production

Among developing countries, Brazil is rather privileged in terms of land resources and technological capacity, and its ethanol program (PNA) cannot be regarded as an example for countries that may have neither one requirement nor the other for large-scale production. Nevertheless, PNA's impact (Chapter 2) may have lessons for other countries.

By the end of PNA's first five-year period, the production of 22.6 m.bbl of anhydrous and hydrous ethanol in 1980 required under 1.0 million ha of harvested land, 207 distilleries of varying scale, and a total investment (excluding land purchases) of about $2000 million at mid-1980 prices—roughly equivalent to an investment of $0.57 per liter. This production had a net balance-of-payments impact of about $520 million in 1980 alone, and created direct employment for about 41,000 permanent workers and 83,000 seasonal workers. This implies investments of about $15,000 per worker or $22,000 per worker-year, which are considerably lower than those required in most energy subsectors, including other renewables and oil.

Against these achievements, however, PNA contributed to a concentration in personal and inter-regional income distribution over the first five-year period. Ethanol investments, which were highly subsidized over this period,[1] were largely and increasingly concentrated in the richer region and the richest state. Although sugar mills and ethanol distilleries were allowed by legislation ("Sugar-Cane Statute") to control up to 40 percent of their raw material requirements, most of them increased this percentage through land purchases, leading to an increase in landownership concentration. In 1981 some distilleries controlled over 80 percent of their requirements, while the national average was about 65 percent. In association with this trend, food staples were increasingly displaced as sugar-cane acreage for ethanol grew sharply after 1977. This implied substitution of large plantations for smaller land plots and, as a result, promoted seasonality and migration of workers as well as eviction of small farmers and subsistence peasants who held land plots under rentals, partnerships, and illicit forms of land tenure.

For the output of about 10,000 m.l in 1985 (2.8 times higher than in 1980 and requiring roughly a threefold increase in investments and harvest land), the net balance-of-payments impact may have been 2–4 times higher, depending on price trends and adjustment in the oil-refining sector. The direct employment effect may have increased by a factor of 2.5–3 as a result of two offsetting factors: increasing mechanization and increasing share (in ethanol output) of regions with relatively high labor coefficients.

The fast growth of ethanol production and imbalances between the supply and consumption of fuels has led to increasing surpluses of petrol and fuel ethanol. While this problem has been solved by exporting petrol and by promoting the consumption of fuel ethanol, the benefits of these short-term solutions are questionable. Fuel ethanol is consumed almost entirely by cars for private use, with limited impacts on growth of output and income. The net benefits of exporting petrol depend on the real economic cost of the petrol surplus. This cost remains uncertain even to energy planners in Brazil and should be assessed in the context of the overall options for increasing the petroleum-substitution impact of ethanol. These options include (a) investments in the oil-refining sector to reduce the output of petrol and increase that of diesel; (b) decreases in crude oil imports coupled with increases of separate imports of LPG, diesel, and possibly other oil derivatives; and (c) increased substitution of ethanol for diesel through changes in engines and transport pattern. The option of reducing the growth rate of ethanol output should also be considered, particularly if (a) the optional adjustments in the oil-refining sector continue to yield a high surplus of petrol and the real economic cost of this surplus is close to its export value, and (b) the substitution of fuel ethanol for diesel in commercial and public-utility vehicles remains constrained by difficulties in changing the relative prices of diesel, gasohol, and fuel hydrous ethanol.

If ethanol production increases in poorer regions and states, the effect on inter-regional income distribution will improve, particularly if small-scale distilleries are given preference under policies that can explore their potentially greater impacts on rural development. This had not happened up until 1984, but these potential benefits deserve more attention than they have received so far. For example, it is technically more viable for small distilleries to make productive use of stillage for irrigation and biogas production than large ones; employment and income benefits can be more decentralized; crop substitutions can be more easily avoided and, as some government-sponsored studies have suggested, cultivation costs are likely to be lower either because of lower transport costs or because intercropping may be more feasible. However, the expected improvement in the effect of production on inter-regional income distribution may or may not contribute to a more equitable distribution of personal incomes, depending on policies concerning land use, land tenure, and regulation of sugar-cane supply. Due to local conditions of climate, soil quality, and land use, the expansion of sugar cane for ethanol in the North-East region is more likely to displace crop land than pastures and promote greater concentration of landownership and incomes than in the southern regions.

Policy options that can improve PNA's impact on both inter-regional and personal income distribution have financial penalties, but this issue lacks research and policy efforts. For example, an increased share of relatively small farmers in the supply of raw material would increase the unit costs of ethanol to the extent that sugar cane from external suppliers means higher raw material cost and risks to the distilleries. Furthermore, financial costs of ethanol production in poorer regions are higher. Thus, a greater decentralization of production through small-scale distilleries would involve higher investments per liter. However, macroeconomic benefits of concentration versus distribution of income have not been fully evaluated in Brazil. Moreover, the benefits of job creation and of additional opportunities for rural settlement have not been adequately taken into account by PNA's policymaking network. Given Brazil's landholding structure, which is characterized by a high degree of marginal forms of land tenure, it is both desirable and possible to create more stable employment in sugar-cane cultivation (the research and training implications of this are discussed in the second section of this chapter, below).

By mid-1985, PNA was being adversely affected by the sharp fall in oil prices. This is beyond the analytical scope of Chapter 2, but some general remarks may be added here. The main policy issues facing the Brazilian government in early 1986 (with spot prices falling to $10/bbl) were whether and how to continue the rate of ethanol output and investments in production; and whether the fixed 0.60–0.65 price parity ratio of ethanol to gasohol should be altered so as to reflect the real costs of these fuels, thus making ethanol more expensive to consumers.

A main argument seen in the media was that ethanol production at an overall economic cost of $50–55 per petrol-equivalent bbl (Pereira, 1986) should be stopped or significantly reduced even if only temporarily. This is nonetheless a naive position with two major flaws. First, more than 90 percent of all cars being produced and purchased since 1983 are ethanol cars, which cannot run on petrol or gasohol. This was a major error, but an irreversible one. Second, Brazil was not buying all of its imported oil in 1985/86 at going spot prices due to some trade agreements. A counter argument discretely pledged by the ethanol lobby was that ethanol production should not be reduced and the policy guidelines should not be altered in ways that would make private production unfeasible. An intermediate and more pragmatic view apparently followed by the government in early 1986 was to maintain current output while freezing new investments, cutting subsidies on existing investments, and allowing ethanol prices to surpass gasohol prices. In spite of these adjustments, the difference in the economic costs of supplying ethanol and petrol were about $23–28 per petrol-equivalent bbl in early 1986. This is a significant cost penalty, but reducing output to levels that would dismantle or wound production capacity and ruin government credibility would be likely to have harsher effects in the near future. The key factor is the world price of oil. This should evidently be the main parameter for new investments in energy supply but it should not determine whether ongoing investments ought to continue or stop. Oil prices will remain unstable and are likely to increase again in response to expected agreements on oil supply as well as additional consumption caused by some economic recovery and lower prices in the 1984–1986 period.

The results of the ex-ante evaluation of two optional production scenarios of 2000 m.l and 3000 m.l of ethanol as substitute for oil fuels and petrochemicals in India (Chapter 3) are very sensitive to variations in critical assumptions about prices of oil products, crop yields, capital costs, social discount rate, input prices, and implementation schedules. This means, of course, that the results must be understood as parameters for further investment analysis rather than final conclusions. This consideration is very important because, on the one hand, assumptions made about the valuation of oil products are the most critical and, on the other hand, world oil prices have been extremely unstable. Thus, among many assumptions regarding the key variables mentioned above, naphtha export prices in 1980–1982 ($305/t, 12 percent higher than crude oil c.i.f prices of $36/bbl) are assumed to change to $300/t (at 1983 prices) in 1990 and 2010.

Based on this price level, and assuming a limited quota on sugar exports and estimated economic costs of producing sugar cane (taking into account opportunity costs of land and other inputs as reflected by their shadow prices), the social cost/benefit analysis concludes that the ethanol program is economic (from the standpoint of society). The sensitivity analysis of changes in key parameters shows that the ethanol program is not economical if benefits are 7 percent lower than those estimated in the reference solution. This means that if world oil prices remain at the levels of 1984–1986, the program is not economic. In 1982, the assumptions made about future prices of naphtha, diesel, and other oil products appeared definitely conservative, but in 1986 the same assumptions imply an overestimation of benefits as a result of the decline in world oil prices in both nominal and real terms. However, this conclusion does not rule out the possibility that a program initiated in the second half of the 1980s will be economic, since oil prices may again increase in real terms during the productive life of investments.

The results of the study are much more sensitive to changes in benefits than to changes in major cost components. For example, if the economic cost of sugar-cane cultivation is 15 percent higher than that estimated in the reference solution, ethanol production is not socially profitable. The profitability of the program is not seriously affected even if capital costs are higher by about 20 percent.

The analysis suggests that the employment and income distribution effects of expanding ethanol production in India are quite significant but this needs further study. Since ethanol is not produced from sugar cane in India, the analysis had to be based on employment data for sugar mills, which are not very reliable. The estimated employment impact for India is much higher than the measured impact for Brazil, and this may reflect not only the sharp differences in the labor markets of the two countries but also a possible overestimation for India. It is also very difficult to determine the effects that sugar-cane ethanol may have on the organization of sugar-cane supply (including the extent to which small farmers may or may not be able to participate in this supply) and on the pattern of landownership.

Given the limitations of data, some of the assumptions and estimates made in the study may be wrong and the sensitivity analysis may not reflect all possible distortions. Nevertheless, the evaluation strongly suggests that the policy decision of not considering ethanol for fuel uses in India should be reviewed once world oil prices increase or stabilize beyond some $20/bbl.

The findings of the country studies of Brazil and India may help research and policy in other countries. A major lesson is that any coherent policy on ethanol production should not be based on general notions about economics of converting sugar cane into ethanol/sugar. As a result of the decrease in oil prices in 1984–1986, Brazil's program has become uneconomic[2] and, at present, it would be unwise for India to decide to expand ethanol production based on sugar cane. Because the oil market has been so unstable,[3] however, it is not a sound policy for any country to rely exclusively on oil price signals as a basis for government investments in energy supply. It would certainly be a wise policy for India to follow up on the preliminary findings of Chapter 3, by promoting research on critical issues identified in that study. Such a follow-up, if well coordinated by the government, could lead to a plan for action in the event that prices or other key economic parameters change favorably in the future.

Most of the literature on ethanol has been biased against the large-scale production of ethanol because major ethanol feedstocks would compete against food crops for farmland. Thus, it has been argued[4] that production would restrain or reduce the growth of food output in countries with scarce land resources. This may, however, be challenged in some situations. The important questions are (a) whether underutilized land would be used for food production anyway; (b) which factors account for increases or decreases of food output in the absence of competing land uses such as ethanol production; and (c) whether this production can help stabilize supply and prices of agricultural commodities like sugar, molasses, manioc starch, and so on. The adequacy of different raw materials and their yields and economic value as food in domestic and external markets may vary from country to country. The amount of arable land required for production depends on the raw materials used and the scale of the programs.

In a survey of six Asian countries, Koide et al. (1982, p. 109) concluded that "at the very lowest achievable prices for cassava ($17.6/t for 22 percent starch) or sugar cane ($16/t), the raw material inputs alone exceed the ex-refinery price of gasoline (petrol) or diesel and the spot prices for the refined products on the world market." This conclusion is misleading, since that study also shows that there are considerable variations in the prices of molasses, sugar cane, and manioc in the six countries studied. For example, the price of molasses in India has been fixed at $7/t as compared with an export price of about $32/t and an Indonesian price of $120/t. Similarly, the cost of sugar cane is said to vary from $17/t in India to $31/t in Thailand. These variations indicate that results based on average figures of costs of production will have limited relevance to actual planning. The present level of costs and prices of sugar cane/manioc may reflect the existing conditions of demand and supply of these products. If assured demand at a remunerative price is guaranteed, there may be considerable scope for higher yields and subsequent reductions in unit costs. As seen in Chapter 3, the costs of production of sugar cane in India were $10–$11/t in 1982 in some regions, which is a lot less than the figure of $16/t, at which the raw material inputs alone were said to "exceed the ex-refinery price of petrol or diesel and the spot prices for the refined products on the world market." Furthermore, transport costs and bottlenecks in developing countries would suggest that comparison should be made in terms of the delivered cost to the user rather than in terms of ex-refinery prices. Distribution costs of hydrous ethanol in particular will normally be lower than those of petrol. Large variations are also found in the prices of petrol and diesel in those six countries, which reflect the role of the governments in fixing these prices independently of import costs or export values.

Even if shortage of petrol is not a major problem in some developing countries (as, for example, in Brazil and the countries reviewed by Koide et al.) a substitution of ethanol for petrol is not necessarily less valuable. Since a reduction in the demand for petrol may enable the refineries to increase the output of naphtha, ethanol can be considered as an indirect substitute for imported naphtha, used mainly by the fertilizer or petrochemical industries. In case there are no imports of naphtha, there is considerable international demand for both petrol and naphtha, which could be exported. The case study of Brazil suggests that there is very limited scope for export-oriented production of ethanol, but international demand is likely to increase, especially if real oil prices begin to rise again.

In summary, the issues relating to ethanol production and the agricultural economies of the developing countries cannot be approached simply in terms of simplistic comparisons between costs of ethanol and ex-refinery prices of petrol. Governments may be concerned with the objectives of promoting employment, regional development, and better income distribution; these cannot be adequately reflected in analyses based on such comparisons.

In planning for ethanol programs based on cultivated crops in other countries, the following points seem to deserve major attention:

- before evaluating choices between ethanol and food production, it is necessary to start with a detailed analysis of fuel supply and consumption trends, taking into account oil-refining capacities where these exist, since, under certain conditions, these can strongly affect the economic significance of ethanol programs in the first place;
- there are clear tradeoffs between cost efficiency and redistributive effects which should be studied carefully before deciding to produce ethanol and establishing program guidelines and production strategies;
- the impacts of farming for ethanol production on income distribution depend primarily on agrarian structure and related legislation; in countries where land is scarce and landlessness is high, the benefits of job creation may not necessarily be significant and overall impacts on income distribution may be negative.

The ex-post assessment of Brazil's experience in ethanol production suggests that ethanol programs need to be planned carefully, particularly with regard to the organization of production, since this is the main determinant of redistributive effects. In Brazil, the reliance on large sugar-cane farms increasingly owned by the distilleries has meant lower ethanol production costs than would have been the case if larger proportions of sugar cane were purchased by distilleries from small farmers. However, this has contributed to an increase in income disparities in rural areas. The analysis for India assumes that sugar cane for ethanol production in India would be supplied by typical small farmers, but while this is more likely to occur in India than in Brazil, it cannot be taken for granted. This would depend on how the government organized sugar-cane supply for ethanol. It is noteworthy that in Brazil, when the ethanol program was launched, landownership in sugar-cane cultivation for ethanol became more concentrated than cultivation for sugar was before 1975.

Comparative Microeconomic Evaluation of Renewable Energy Technologies for Irrigation Pumping

The financial and economic feasibility of different water-pumping technologies for irrigation depend not only on technical performance, capital costs, and other basic economic parameters, but also on natural resource conditions, cropping patterns and farm size, fuel pricing policies, and availability of technical capacities for operation and maintenance. Since comparative evaluations and ranking of different technologies must be based on certain assumptions about these variables, and about inflation, fuel-price escalation rates, and capacity utilization rates, it is possible to prove that a certain technology is better than another for users and/or society. This qualification is necessary for interpreting the findings of Chapters 4, 5, and 6, among other literature on the subject.

For Egypt and Bangladesh (Chapter 4), the sum of the present values of capital and operating costs of human-, animal-, and diesel-powered pumps for low-lift irrigation are compared at mid-1979 prices. The economics of traditional energy sources depend crucially on the farm size in question and (partly in relation to this) on the opportunity cost of labor and the real resource costs of maintaining animals, which are determined by the ways in which animals are utilized. Assuming world prices of diesel in mid-1979, diesel engines are economic only for large farms that do not maintain animals for other purposes.

Estimated annual unit costs in $/1000 m^3 are used as a basis for comparison. The study concludes that the human-powered lifts are the most attractive option for typical small farms in those two countries where the farmers' own labor is used. For medium farms (3.1–4.6 ha), animal-operated lifts are the most attractive, even when full capital costs of purchasing of costs are taken into account. Due to capacity underutilization, the present values of costs of the diesel pumpset (using a 3.8 kW engine) are higher than those of animal-powered lifts in the two countries. For large farms, however, the diesel pumpset is more economic because capacity utilization is higher, whereas costs of hired labor and animal feeding are too high. The costs of alternative technologies are higher because, for a large farm, it is necessary to hire labor for irrigation. In the case of animal-operated devices, four units will be required, thus increasing the costs proportionately. Under these assumptions, a diesel pumpset is more economic than a traditional system. The only case where an animal-powered lift can compete with a diesel pumpset is when the capital costs of cows are estimated at real resource costs (of rearing cows) instead of using market prices.

In Chapter 5, windmills are financially and economically compared against at least one other alternative technology in terms of total unit costs per m^3 of water output at 1980/81 prices. The conclusions cannot be generalized, since the chapter analyzes four country situations under slightly different methodologies. Windmills are compared with a kerosene pumpset in Sri Lanka, with a diesel pumpset in Kenya and the Sudan, and with diesel, solar-photovoltaic, and solar-thermal systems in Cape Verde.

Based on the assumptions of the studies, the financial unit costs of windmills are lower than those of the alternative technologies considered in all the four countries, except in Sri Lanka. At world prices rather than market prices of fuels used by the competing technologies, the windmill is economically feasible in Sri Lanka and the Sudan.

The unit cost of water pumped by windmills varies from one country to another. For example, in Cape Verde, the difference between the unit cost of windmills and that of the option considered ranges from 2 percent in the ribeiras (lowlands) to 76 percent in the achadas (highlands), due to differences in wind speeds or pumping head. The cost variation is about 34 percent in Sri Lanka when the economic cost of kerosene is taken into account. In Kenya, the cost difference is 82 percent in favor of windmills in Lower Tana, which is the region with medium wind speeds, and about 33 percent in Kisumu, the region with relatively low wind speeds. In the Sudan, the cost advantage of the windmills ranges from 7 to 16 percent, depending on the cost of diesel oil. This shows that in some regions and sites, windmills can be feasible, both financially and economically. However, these cost estimates are based on a number of assumptions regarding wind speed, availability of water, capacity utilization of windmills/diesel pumpsets, and so on, which must be taken into account.

The four country studies thus show that the economics of windmills are very location-specific, since water output depends crucially on groundwater availability and wind speed. The water output and unit costs are sensitive to the total pumping head, as illustrated by the case study of Cape Verde. The total costs and the unit cost (per m^3) of a windmill can be significantly affected by the cost of a storage tank because of variations in wind and water supply. The unit cost depends mainly on the assumptions made about the capacity utilization of windmills and alternative devices. However, the total quantity of water available may not be utilized for a variety of reasons. For example, farm size may be smaller than necessary to utilize full water output; cropping pattern may be such that water requirements are seasonal and pumped water cannot be fully utilized in the off-season; there may be a mismatch between the monthly requirements and water output from a windmill because of cropping patterns, wind speeds in different months, and so on. Under these conditions, a comparison of unit costs of water pumped can lead to misleading conclusions. One alternative method is to compare the unit costs for a given farm size and cropping pattern. For example, in Kenya, it is shown that if water availability in the wells restricts the area irrigated to 0.5 ha, the windmill becomes an attractive proposition. In this case, the unit cost of the windmill is about 60 percent lower than that of a diesel pumpset, because the capacity of the diesel option will be highly underutilized. Even if water is available, farm size may nevertheless constrain capacity utilization. For small farms, it becomes important to consider smaller windmills, petrol engines, and photovoltaic systems.

The four country studies suggest that windmills look more attractive economically if foreign exchange outflows and employment generation are taken into account. For example, in Sri Lanka, annual foreign currency expenditures are estimated at $162 in the case of a windmill and $292 in the case of a kerosene pumpset. In Kenya, a small windmill replacing a diesel pumpset is expected to provide annual foreign exchange savings of about $550 in capital cost instalments and about $175 in fuel costs (all figures at 1980/81 prices). Regarding employment, 16 worker-days of labor are utilized just in the installation of a windmill in Sri Lanka. In Kenya, windmill production in 1982 employed more than 35 persons full-time. In Cape Verde and the Sudan, the extra employment created by the local manufacture, sale, and installation of a windmill is estimated at 70 and 60 worker-days, respectively.

Overall, the country studies summarized in Chapter 5 suggest that windmills for water pumping can be economically feasible under favorable conditions of (i) wind speed and frequency and (ii) groundwater supply. In some cases, the economic feasibility of the windmills is constrained by subsidies on oil-derived fuels. Thus, if oil-derived fuels need to be subsidized, governments should also partly subsidize the capital costs of windmills, since these costs are higher than those of conventional technologies and are a major constraint to the use of windmills (this also applies to some other options analyzed in Chapter 6).

While not necessarily altering the above conclusions, the basic assumption underlying the cost comparisons in Chapter 5 (e.g., that different options are equivalent in terms of water output and benefits) have some limitations, as discussed in the introduction to Part II. The methodology used in Chapter 6 for an ex-ante evaluation of diesel, biogas, wind, electricity, solar-photovoltaic, and thermal systems for pumping tries to overcome the methodological limitations of the previous chapters and other literature on irrigation.

In this case study of northern India, the capacities of the optional technologies are selected in such a way that all of them can yield the same levels of water output and irrigation benefits for certain types of farming conditions. The technologies are then compared in terms of their cost-effectiveness, measured by the sum of the present values of capital costs and operating costs incurred over a fixed period.

The study finds that the use of electricity from the regional grid for irrigation pumping is consistently more expensive from the viewpoint of society than diesel and diesel-cum-biogas. Use of biogas in a dual-fuel engine is found economic under a variety of assumptions. Producer gas used in a dual-fuel engine is less attractive than biogas but it can be economic where costs of providing electricity/diesel are high and the economic costs of fuelwood production and use are low. The use of a solar-photovoltaic pumping system is economic for the small farm of 0.4 ha, even under current (1984) costs and current efficiencies for regions/farms where the capital costs of providing grid electricity are higher than Rs 25,000 per pumpset of 4 kW. Thus, solar-photovoltaic systems may be considered as a viable alternative for areas that are away from the grid and have low load factors of demand. Solar-thermal systems become economic when significant improvements in the technical efficiency of utilization of solar energy can be made. Windmills are two to three times more expensive than conventional systems because wind regimes at Ghazipur are not suitable. This underlines the fact that the economics of windmills are site-specific and need detailed analysis for a number of windy locations before a program for their promotion is launched.

The findings show that there is a need for a more selective approach in rural electrification that can thoroughly assess estimation and utilization of load, distance from a grid, and costs of alternative sources. Since only a few dual-fuel engines were being used in India in 1984–1986, it is important to analyze institutional and organizational aspects in detail. In particular, policies of pricing and tax/subsidies on energy sources must be studied since low prices of electricity and/or diesel result in low cash savings from the use of renewable energy technologies, thus making the latter financially unprofitable. It is also important to emphasize the employment and income distribution effects of using biomass resources for irrigation.

RESEARCH AND TRAINING IMPLICATIONS

The analysis of ethanol production in Brazil shows the following major gaps in research and training:

- strategies and organization of ethanol-related agriculture;

- training and management of labor;
- use of distillery by-products.

Concerning the first of these, a main issue is to promote the combined use of feedstocks in order to increase the annual period of operation from 150–180 days to about 300 days. This can be achieved by using sweet sorghum during the off-season of sugar cane and does not imply any major additional investment in distillery equipment. Research in Brazil on this issue has reached acceptable results but no such consortia were being used on a commercial scale by 1985. The combined use of sugar cane with manioc roots or chips is also promising but more constrained than the other consortium because an additional investment of 15–20 percent in distillery equipment is required and also because manioc cultivation on a commercial scale is not well developed in Brazil. The potential benefits of consorting raw materials are an increase of financial returns on investment and both an increase and a more stable demand for labor year-round.

Regarding training and management of labor, a key issue is not only to improve workers' skills for specific job descriptions (which has been promoted by PLANALSUCAR and private organizations) but also to enable part of the workforce to take different jobs during the harvest season and the off-season, so as to increase year-round employment. This appears especially important in areas where there are a large number of landless migrant workers.[5]

Regarding the use of ethanol by-products—stillage and bagasse—there is a need to explore ways of optimizing the use of stillage in irrigation by better use of field channels and pipelines and promoting production of briquettes from surplus bagasse for use in industries located near sugar-cane distilleries. Biogas can be produced from stillage,[6] but the field experience has been limited to a few pilot experiments and only one commercial project in the state of Sao Paulo. The possibility of reducing costs of this production through technological innovations needs further research, but both the benefits of such research and the actual adoption of biodigestion facilities by distilleries depends on the creation of demand for biogas in industries and communities nearby.

It is crucial to assess the economic cost of petrol exports, taking into account different oil-refining scenarios and relative border prices of different oil products and implications of such cost for ethanol production strategy. Also important are the technological, economic, financial, and social implications of possible decentralization of future ethanol production through greater use of small-scale distilleries (assuming that oil prices will increase to their 1980–1984 levels). So far, these aspects have been studied separately and no policy conclusions have emerged.

The ex-ante case study of ethanol programs in India shows the need for better data regarding capital costs, crop cultivation costs, patterns of land use, crop substitution, employment in sugar mills and distilleries, sugarcane transport, and crop cultivation in general. There is also a need for detailed follow-up studies regarding fluctuations in export prices of sugar, fluctuation in cane output, impact on crop substitutions, employment creation, and the explicit premium on domestic substitutes for imported oil. These needs can be disaggregated into a much larger list of research topics, but it is pointless to carry out research on this issue in India until the government takes a more open position toward ethanol as an option and, of course, until oil prices regain their upward trend.

The findings of Part II also show a number of priorities for research, training, and technical cooperation. In particular there is a need for detailed comparative economic evaluations of alternative technologies for given end-uses and locations under a variety of assumptions, regarding cropping patterns, farm size, availability of groundwater, availability of local resources, and seasonal variations in solar radiation or wind speeds.

Although some information on field testing of renewable energy devices is available, the data need to be strengthened by a proper evaluation of these systems. It may be necessary to provide for a built-in mechanism of simultaneous and ex-post evaluation as a part of each project on renewable technologies. In addition, it is desirable to have an apex body that can coordinate the activities of various organizations connected with R & D, field-testing, manufacturing, and financing of renewable energy technologies. This apex body may:

- determine the optimum mix of energy alternatives from the viewpoint of society;
- ascertain policy changes that would be required in energy pricing and tax/subsidy policies in order to make these technologies financially viable for the users; and
- indicate organizational requirements to promote the diffusion of these technologies.

For that coordination to be effective, however, it is necessary to improve the availability and reliability of data on the technical performance of different renewable energy options for irrigation and other end-uses.

As a result of the recent decline in world oil prices, some technologies that were available practically off the shelf in 1979–1983 may have been discontinued or their capital costs may have increased. Will such technologies be available in 1986–1990; and if they are not, what will be the costs of recuperating their readiness? This issue may deserve some research as a basis for economic evaluations of technologies at present or in the future.

Another important set of issues that deserve attention in follow-up to the socioeconomic analyses presented in this book concerns the organizational and institutional aspects of the commercial diffusion of renewable energy technologies for irrigation pumping and other end-uses. The findings of Part II showed that, for given conditions, some of those technologies were economically more attractive than conventional energy options over the 1980–1984 period when the evaluations were carried out. Yet the commercial diffusion of dual-fuel engines, windmills, or photovoltaic systems has been limited. What are the explanations for this? Perhaps the expectations were too high and the lags in the adjustment to higher prices of conventional energy sources were longer than could have been anticipated. Perhaps part of the problem lies in the means of developing, introducing, producing, and commercializing some of the technologies. What have been the organizational and institutional barriers in the diffusion process and which of these can be removed? These are questions for the future.

NOTES

1. Interest on government loans increased after 1981 but remained subsidized until 1986.
2. The economic cost of ethanol was lower than the estimated economic cost of petrol between 1979 and 1982. For details, see Pereira (1986, Chapter 3).
3. For a comprehensive discussion of recent literature on oil market trends and their general policy implications, see Pereira, Ulph, and Tims (1987).
4. See, for example, Rask (1979) and FAO (1980).
5. For a more comprehensive discussion, see Pereira (1986) and Graziano da Silva (1980).
6. Each liter of ethanol produced from sugar cane also results in about 12 liters of liquid waste, rich in organic materials and nutrients. In Brazil, a small part of this effluent is used for direct irrigation and the rest is treated in reservoirs and later discharged into rivers and lakes. In some cases, serious water pollution problems have occurred because of noncompliance with established procedures of effluent treatment. Since sugar-cane distilleries use bagasse for their energy needs, it would be worth exploring possibilities of piping surplus effluent to biodigestors in rural communities and industries nearby to serve household and/or productive needs while avoiding pollution effects. One commercial project was underway in a distillery in the state of Sao Paulo in 1987 to provide biogas to an industry near the distillery.

Bibliography

Agricultural Refinance and Development Corporation (ARDC) (1980). *Selection of a pumpset* (New Delhi: ARDC).

Agroanalysis (1979). "Proalcool" (Rio de Janeiro: Fundaçao Getúlio Vargas), Vol. 2, n. 11.

Asset (various issues; Tokyo, United Nations University).

Auer, P. (Ed.) (1981). *Energy and the Developing Nations,* proceedings of Electric Power Research Institute Workshop, Hoover Institution, Stanford University, March 18–20, 1980 (New York: Pergamon Press).

Bajracharya, Deepak (1983). *Rural energy planning in the developing countries of Asia.* Paper presented at FAO/ESCAP seminar on Rural Energy Planning, Honolulu, Resource System Institute, East–West Center, 9–29 April.

Barnett, A., et al. (1978). *Biogas technology for the Third World* (Ottawa: IDRC).

Beurskens, H.J. (1981). *Wind energy for water pumping in Cape Verde* (Amersfoort, the Netherlands: Consultancy Services Wind Energy Developing Countries—CWD; mimeographed; Feb.).

Bhalla, A., D. James, and Y. Stevens (eds.) (1984). *Blending of new and traditional technologies; case studies* (Dublin: Tycooly International Publishers Ltd.).

Bhatia, R. (1977). "Economic appraisal of biogas units in India: Framework for social benefit-cost analysis," in *Economic and Political Weekly* (Bombay: Sameeksha Trust), Vol. XII.

______ (1980a). "Economics of energy alternative for irrigation pumping: Some results for small farms in North Bihar", in Pachauri, R.K. (Ed.), *International Energy Studies* (New Delhi: Wiley Eastern).

______(1980b). *Energy and rural development—An analytical framework for socio-*

economic assessment of technological and political alternatives (Honolulu: Resource System Institute, East–West Center; mimeographed; Jan.)
______(1982). "Technological problems of the oil industry in India" (New Delhi: Institute of Economic Growth; mimeographed; Dec.).
______(1983a). *Planning for ethanol production in developing countries: An analytical framework and a case study for India* (Geneva: ILO, Technology and Employment Branch; unpublished; Feb.).
______(1983b). *Planning for the petroleum and fertiliser industries: A programming model for India* (New Delhi: Oxford University Press).
______(1983c). *Infrastructural investments for rural energy: A review of evidence for India* (New Delhi: Institute of Economic Growth; draft paper prepared for Asian Development Bank).
______(1985). "Energy pricing in developing countries: an analysis of investment allocation and consumer choices," in Siddayao, C. (Ed.), *Criteria for Energy Pricing Policy* (London: Graham & Trotman, forthcoming).
Bhatia, R., and M. Mehta (1975). "Tubewell irrigation: An analysis of some technical alternatives," in *Economic and Political Weekly* (Bombay: Sameeksha Trust; Dec.).
Bhatia, R., and M. Niamir (1979). *Renewable energy sources: The community biogas plant* (Cambridge, MA: Harvard University; mimeographed).
Birch, D.R., and J.R. Rydzewski (1980). *Energy options for low-lift irrigation pumps in developing countries* (Geneva: ILO; mimeographed World Employment Programme working paper n. 57; restricted).
Bose, J. (1982). *Impact of electrification on Indian agriculture with special reference to Bihar* (New Delhi: Indian Agricultural Research Institute; mimeographed).
Bremen Overseas Development Agency (BORDA) (1981). *Biogas in the People's Republic of China* (Bremen; mimeographed summary of project report).
______(1983). *Report of the workshop on uniformity of information reporting on biomethanation systems* (Washington, DC: Equity Policy Center).
Brown, N.L., and J.W. Howe (1978). "Solar energy for village development," in *Science* (Washington, DC: American Association for the Advancement of Science), Vol. 199, Feb. 10.
Buch-Larsen, B. (1979). *Economics of windmills in Sri Lanka: Four case studies* (Colombo: Wind Energy Unit, Water Resources Board).
CENAL (1981). *Novas condiçoes de financiamento do PROALCOOL* (Brasilia: MIC).
______(1983). *Avaliaçao do programa nacional do àlcool, 1981/82* (Brasilia: MIC; mimeographed).
Central Power Research Institute (CPRI) (1979). *Report of the economic viability of water pumping windmills for rural applications -The WP2 windmill* (Bangalore: Technical Report n. 95).
Charlu, T.G., and D.K. Dutt (1982). *Ground water development in India* (New Delhi: Rural Electrification Corporation).
Chetty, W.K. (1981). *Economics of price and distribution controls—I: The sugar industry.* Paper presented at the Meeting of the Econometric Society, Pune, India, Jan.
CNPq (1973). *Evoluçao tecnológica do álcool etílico* (Brasilia).
COALBRA (1983). *Alcool e emprego: o impacto da produçao de álcool de cana e de madeira na geraçao de empregos* (Brasilia).

Conjuntura Econômica (various issues) (Rio de Janeiro: Fundaçao Getúlio Vargas).
COPERSUCAR (1981). *Aspectos econômicos da produçao de cana, açucar, e álcool—período 1978–80* (Sao Paulo: COPERSUCAR; mimeographed).
______(1983). *Atualizaçào dos custos de produçao de cana, açucar e álcool—safra 1982–83* (Sao Paulo: COPERSUCAR; mimeographed).
Dastane, N.G. (undated). *A practical manual for water use research* (New Delhi: Indian Agricultural Research Institute; mimeographed).
Datta, R., and G.S. Dutt (1981). "Producer gas engines in villages in less developed countries", in *Science* (Washington, DC: American Association for Advancement of Science), April.
deLucia, R., and R. Bhatia (1980). *Economic evaluation of renewable energy technologies: A review* (Washington, DC: National Academy of Sciences; mimeographed).
deLucia, R., and M. Lesser (1980). *Energy from organic residues in developing countries: Present use, potential and observation on feasibility,* Paper presented at the Bioenergy World Congress and Exposition, Atlanta, GA, Apr.
Dhawan, B.D. (1977). "Tubewell irrigation in the Gangetic plains," in *Economic and Political Weekly* (Bombay: Sameeksha Trust), Sept.
Dunkerley, J., W. Ramsay, L. Gordon, and E. Cecelski (1981). *Energy strategies for developing nations* (Baltimore: Johns Hopkins Univ. Press for Resources for the Future).
Dutt, G.S. (1978). "Reducing cooking energy use in rural India" (Princeton, NJ: Princeton University, Center for Environmental Studies; mimeographed; Nov.).
El-Hinnawi, E., X. Biswas, R. Margaret, and A. Biswas (Eds.) (1983): *New and Renewable Sources of Energy* (Dublin: Tycooly International Publishing Limited).
ESCAP (1981). "Renewable sources of energy," in *Wind Energy* (Bangkok: Economic and Technical Co-operation among Developing Countries), Vol. III.
______(1980). "Renewable sources of energy," in *Solar Energy* (Bangkok: Economic and Technical Co-operation among Developing Countries), Vol. III.
______(1979. "Proceedings of the workshop on biogas and other rural energy resources and the roving seminar on rural energy development," in *Energy Resources Development* (New York: United Nations), series n. 19.
______(1982). "Small hydro-power development," in *Renewable Sources of Energy* (Bangkok: Economic and Technical Co-operation among Developing Countries), Vol. IV.
FAO (1974). *Prospective study of agricultural development in Bangladesh* (Rome; mimeographed).
______(1977). *Crop water requirements, irrigation and drainage* (Rome: FAO; mimeographed paper n. 24).
______(1981). "Energy cropping versus food production," in *Agricultural Services Bulletin* (Rome; mimeographed paper n. 46).
FAO/World Bank (1978). *Cooperative programme report on the Bangladesh low-lift pump project preparation mission* (Rome; mimeographed; Oct.).
Farinelli, Ugo (1983). *A brief review of solar photovoltaic energy conversion* (Paris: Organisation for Economic Co-operation and Development; mimeographed; Apr.).
Finlay, John (Ed.) (1982). *Guidebook on biogas development* (Bangkok; ESCAP).
Floor, W.M. (1978). *Energy options in rural areas of the Third World.* Paper presented for

the Eighth World Forestry Congress, Jakarta.

Fluitman, F. (1983). *The socio-economic impact of rural electrification in developing countries: A review of evidence* (Geneva: ILO; mimeographed World Employment Programme working paper n. 126).

Foley, G., and G. Bernard (1983). *Biomass gasification in developing countries* (London: Earthscan).

______(1984). *Farm and community forestry* (London: Earthscan).

Foley, G., G. Bernard, and L. Timberlake (1983). *Gasifiers, fuel for siege economies* (London: Earthscan).

French, D. (1979). *The economics of renewable energy systems for developing countries* (Washington, DC; USAID; mimeographed; Jan.).

Geller, H.S. (1984). *Ethanol from sugar-cane in Brazil: an investigation of some critical issues* (Sao Paulo: CESP; mimeographed).

Gellerson, M. (1979). "Marginal cost-based electricity tariffs: Theory and case-study of India," in *Indian Economic Review* (New Delhi: Asia Publishing House), Oct.

Ghate, P.B. (1980). "Irrigation for very small farmers: Appropriate technology or appropriate organisation", in *Economic and Political Weekly* (Bombay: Sameeksha Trust), Feb.

______(1979). "Biogas: A decentralised energy system," in *Economic and Political Weekly* (Bombay: Sameeksha Trust), July.

Ghirardi, A.G. (1983). *Alcohol fuels from biomass in Brazil: a comparative assessment of methanol and ethanol* (Berkeley: University of California, Energy and Resources Programme; mimeographed Ph.D. dissertation).

Gittinger, J.P. (1982). *Economic analysis of agricultural projects* (Baltimore: Johns Hopkins University Press, 2nd edition).

Goodman, D., and L. Ratcliff (1977). "The bóias-frias: rural proletarianisation and rural marginality in Brazil," in *International Journal of Urban and Regional Research* (London: Edward Arnold Publ.), Vol. 1, n. 2.

Government of Bangladesh (1978). *Statistical year book* (Dacca: Central Agency for Public Mobilisation and Statistics), July.

Government of India (1980). *Report of the Committee of Technical Experts on alcohol and alcohol-based industries* (New Delhi), Jan.

______ (undated). "Studies relating to cost of production of principal crops" (New Delhi, unpublished).

______ (1975). *Report of the Sugar Enquiry Commission* (New Delhi), Oct.

______ (1977). *Report of the Committee to conduct study and make recommendations for the development of industries based on ethyl alcohol* (New Delhi), Dec.

______ (1981). *Economic survey: 1980–81* (New Delhi).

______ (1981). *Draft Sixth Five-Year Plan, 1980–85* (New Delhi: Planning Commission).

______ (1981). *National paper India* (Nairobi: United Nations Conference on New and Renewable Sources of Energy; Aug.).

______ (1983). Planning Commission: *Report of the plan evaluation organisation on rural electrification* (New Delhi).

______ (1983–84). *Report of the Department of Non-Conventional Sources of Energy.*

Government of Papua New Guinea (Department of Minerals and Energy, Energy Planning Unit) (1980). *The Brazilian experience: Its relevance to Papua New*

Guinea and countries with similar circumstances (Washington, DC: Appropriate Technology International; mimeographed report n. 2).

Government of Uttar Pradesh (1980). *Draft Sixth Five-Year Plan, 1980–85* (Lucknow: Planning Department).

Graziano da Silva, J.F. (1980). *O 'bóia - fria': entre aspas e com os pingos nos is* (Campinas, SP: UNICAMP; mimeographed).

______ (1978). *Estrutura agrária e produçao de subsistência na agricultura brasileira* (Sao Paulo: HUCITEC).

Griffin, Keith (1974). *The political economy of agrarian change* (London: Macmillan).

Gupta, S.C., and A. Majid (1965). *Producers' response to changes in price and marketing policies: A case-study of sugar-cane and paddy in Uttar Pradesh* (London: Asia Publishing House).

Halcrow, William, and Partners (1981, 1983). *Small-scale solar-powered irrigation pumping systems, Phase I* (Washington, DC: UNDP/World Bank; mimeographed; June).

Halcrow, William, and Intermediate Technology Power Ltd. (1984). *Handbook on solar water pumping* (Washington, DC: UNDP/World Bank; mimeographed; Feb.).

Hamid, Y.H., and W.A. Jansen (1980). *Wind energy in the Sudan* (Amersfoort, the Netherlands: CWD; mimeographed; July).

Harrison, J., A. Hitchings, and J. Wall (1981). *India: demand and supply prospects for agriculture* (Washington, DC: World Bank; mimeographed World Bank Staff working paper n. 500).

Harriss, B. (1977). "Rural electrification and the diffusion of electric water-lifting technology in North Arcot District, India," in B.H. Farmer (Ed.), *Green Revolution* (London: Macmillan).

Hellinger, D., et al. (1983). *Building local capacity for sustainable development* (Washington, DC: Appropriate Technology International; mimeographed).

Herrera, Amilcar (1979). *The generation and dissemination of appropriate technologies in developing countries: A methodological approach* (Geneva: ILO; mimeographed World Employment Programme working paper n. 51; restricted).

Hurst, C. (1983). *Energy for rural development: A case-study in Bihar, India* (Cambridge, MA: Harvard University; mimeographed M.A. thesis; May).

______ (1984). "Model of an Indian village: A study of alternative sources of energy for irrigation," in *World Development* (Oxford: Pergamon Press), Vol. 12, n. 2.

Hurst, C., and P. Rogers (1982). *Micro-gasification for small-scale irrigation in developing countries.* Paper presented at the First International Producer Gas Conference, Colombo; Marga Institute; Nov.

International Development Research Council (IDRC) (1986). *Energy research—directions and issues for the developing countries* (Ottawa: mimeographed; restricted).

IBGE (1971, 1975, 1977, and 1981). *Anuário estatístico do Brasil* (Rio de Janeiro).

______ (1983): *Levantamento sistemático da produçao agrícola* (Rio de Janeiro).

Indian Agricultural Research Institute (IARI) (1980): *Report on field evaluation and testing of solar pumps* (New Delhi).

Indian Council of Agricultural Research (ICAR) (1978): *Report on biogas* (New

Delhi).
______ (1979). *Report of the Commission of sugar-cane into ethanol* (New Delhi).
Indian Sugar Mills Association (1980–81). *Indian sugar year book* (New Delhi, Vol. I).
Indian Sugar (various issues). (New Delhi: Indian Sugar Mills Association).
Institute of Energy Economics (1981). *The alcohol fuels from energy farming* (Tokyo), Vol. II.
Islam, S. (1980). *Oil imports, foreign exchange and the Brazilian Alcohol Program: Some preliminary observations* (Washington, DC: Resources for the Future; mimeographed working paper; Oct.).
Joshi, D., D. Seckler, and B.C. Jain (1983). *Social forestry, wood gasifiers and lift irrigation: Synergistic relations between technology and natural resources in rural India* (New Delhi: Ford Foundation; mimeographed, Jan.).
Kishore, V.V., et al. (1983). *A solar water pump for lift irrigation,* Report of the Indo-Swiss Collaborative Project (Bhavnagar, India: Central Salt and Marine Chemicals Research Institute; Dec).
Koide, S., R. Brooks, and T. Vicharangsan (1982). *Regional study on production of fuel ethanol from agro-products* (Bangkok: Economic and Social Commission for Asia and the Pacific [ESCAP]; June).
Kumar, C. (1970). "Farmers' response to price changes: Sugar-cane in UP, 1951–52 to 1966–67," in *Economic and Political Weekly* (Bombay: Sameeksha Trust), May.
Leach, Gerald (1979). *Energy.* Report prepared for the Conference "Agricultural Production: Research and Development Strategies for the 1980s," Bonn; Oct.
Lichtman, Rob (1984a). *Manpower and training aspects of introducing biogas and other energy systems in rural areas: A study of the Indian experience* (Geneva: ILO; EMP-POP/PLAN, unpublished).
______ (1984b). *Socio-economic and organisational aspects of biogas programmes: A study of India's experience* (Geneva: ILO; EMP-TEC, unpublished).
______ (1982). *Biogas systems in India* (Washington, DC: VITA).
______ (1983). *US R and D in biogas systems: Applications for developing countries.* Paper prepared for the United Nations Financing System for Science and Technology for Development/UNDP, New York.
Lierop, W.E. van, and L.R. van Veldhuizen (1982). *Wind energy development in Kenya* (Amersfoort, the Netherlands: CWD; mimeographed; Nov.).
Little, I., and J. Mirrlees (1974). *Project appraisal and planning for developing countries* (London: Heinemann Educational Books).
Mackillop, A. (1983). *Technology, employment and development implications of new and renewable sources of energy* (Geneva: ILO; mimeographed World Employment Programme working paper n. 114; restricted).
Madhavan, M.C. (1972). "Acreage response of Indian farmers: A case-study of Tamil Nadu," in *Indian Journal of Agricultural Economics* (Bombay: Indian Society of Agricultural Economics), Jan.–Mar.
Mahdi, S.S. (1982). "Biogas programme in India," in *Financing Agriculture,* Vol. XIV, n. 2–3 (Bombay: Agricultural Finance Corporation Ltd.).
Makhijani, A. (1977). "Energy policy for rural India," in *Economic and Political Weekly* (Bombay: Sameeksha Trust), Vol. XXII, special n. 33 and 34.

Makhijani, A., and A. Poole (1976). *Energy and agriculture in the Third World* (Boston, MA: Ballinger).

Manassah, Jamal T. (Ed.) (1981). *Alternative energy sources,* Part A (New York: Academic Press).

Mani, A., and S. Rangarajan (1981). *An integrated study of solar and wind power availability in India* (New Delhi: National Solar Energy Convention; mimeographed).

Mann, H.S., H.P. Garg, and K.P. Thanvi (1980). "Utilisation of solar, wind and biogas energies in a rural energy centre for arid zone of India—A total energy concept", in Pachauri, R.K. (Ed.), *International energy studies* (New Delhi: Wiley Eastern Ltd.).

Mathur, P.S., and C. Belhari (1980). *Growing sugar-cane in different states in India* (Government of India, Directorate of Sugar-Cane Development, Ministry of Agriculture).

Matlin, R.W. (1981). *Technical aspects of solar pumping systems.* Paper presented at Workshop on Solar Pumping in Developing Countries, Manila, 22–26 June.

Meckhof, R.L., W.E. Turner, and F.D. Holland (1980). "U.S. agricultural policy and gasohol," in *American Journal of Agricultural Economics* (Ames, Iowa: American Agricultural Economics Association), Aug.

Melo, F. Homem de (1980). "Agricultura nos anos 80: perspectivas e conflitos entre objetivos de política", in *Estudos Econômicos* (Rio de Janeiro: Confederaçao Nacional de Indústria), Vol. 10, n. 2.

______ (1985). The Brazilian energy policy and the alcohol program. Paper prepared for the 1985 International Roundtable on Renewable Energy, Sao Paulo (USP, Department of Economics).

Melo, F. Homem de, and E.R. Pelin (1983). *Política agrícola e composiçao da produçao* (Sao Paulo: University of Sao Paulo, Instituto de Pesquisa Econômica; mimeographed).

Ministério de Minas e Energia (MME) (1980 and 1983). *Balanço energético nacional* (Brasília).

Montes, H., and L.F. Gomes (1978). "Expansao ótima dos centros de mistura de álcool com gasolina," in *Brasil Açucareiro* (Rio de Janeiro: IAA), Dec.

Moulik, T.K. (1982). "A case study of community biogas plant," in *Biogas energy in India* (Ahmedabad: R.N. Shah Academic Book Centre), Aug.

______ (1984). *Case study: India's experiment with community biogas plants,* in Bremen Overseas Research and Development Association: *Biogas workshop on community plants* (Bremen: Federal Republic of Germany, proceedings of international biogas workshop on community plants; May).

______ (1985). *The biogas program in India and China: a comparative analysis of experiences* (Ahmedabad: Indian Institute of Management; mimeographed working paper, n. 555; restricted).

Moulik, T.K., et al. (1978). *Biogas systems in India: A socio-economic evaluation* (Ahmedabad: Indian Institute of Management; mimeographed).

Munasinghe, M. (1979). *The economics of power system reliability and planning: Theory and case study* (Baltimore: The Johns Hopkins University Press).

Munasinghe, M., and G. Schramm (1983): *Energy economics, demand management and conservation policy* (New York: van Nostrand, Reinhold and Co.).

Murthy, B.S. (1980). "Diesel engines and alcohol," in *Urja* (New Delhi: Dipak Chaudhuri Editor and Publisher), Feb.

Narain, D. (1965). *The impact of price movements on areas under selected crops* (Cambridge, MA: Cambridge University Press).

National Academy of Sciences (1976). *Energy for rural development* (Washington, DC).

O'Keefe, P. (1983). "Fuel ethanol: unrepeatable Brazil," in *New Scientist* (London: IPC Magazines Ltd.), July 28.

Pandit, G. Patil (1980). *Water pumping in developing countries* (Washington, DC: Systems Consultants Inc.; mimeographed; Oct.).

Pathak, B., et al. (1980). *Studies on sugar-cane as an energy crop for Punjab* (Ludhiana: Punjab Agricultural University).

Peek, P., and G. Standing (Eds) (1982). *State policies and migration: studies of Latin America and the Caribbean* (London: Croom Helm, for the ILO).

Pelin, E.L. (1980). *The impact of Brazil's Proalcool Program on land prices and crop substitutions.* Paper prepared for the Fourth International Symposium on Alcohol Fuels Technology, Guarujá, SP, Brazil, Oct. (University of Sao Paulo, Instituto de Economia Agrícola).

Pereira, Armand. F. (1983). "Employment implications of ethanol in Brazil," in *International Labour Review* (Geneva: ILO) Vol. 122, n. 1, Jan.–Mar.

______ (1986). *Ethanol, employment and development: lessons from Brazil* (Geneva: ILO).

Pereira, A., A. Ulph, and W. Tims (1987). *Socio-economic and policy implications of energy price increases* (London: Gower Publication Co., for the ILO).

Petroleum Economist (various issues). (London: Petroleum Press Bureau Ltd).

PLANALSUCAR, USP, and Centro de Tecnológia Mauá (1981). *Previsao e análise tecnológica do PROALCOOL* (Piracicaba, SP, PLANALSUCAR; mimeographed; May).

Rahmen, B.A. (1984). "The potential of solar cells," in *Petroleum Economist* (London: Petroleum Press Bureau Ltd.), May.

Ramsay, W. (1982). *Subsidies for fuel alcohols in the United States* (Washington: Resources for the Future; mimeographed discussion paper D-73G).

Ramsay, W., and J. Jankowski, Jr. (1982). "Alcohol fuels and the agricultural sector" (Washington: Resources for the Future; mimeographed discussion paper D-73J).

Rangaswamy, P. (1978). *Supply response of sugar-cane and its products in UP* (New Delhi: Agricultural Economics Research Centre).

Rask, N. (1979). *Using agricultural resources to produce food or fuel-policy intervention or market choice* (Columbus, Ohio: Ohio State Univ., Department of Agricultural Economics and Rural Sociology; mimeographed).

______ (1980). *Using agricultural resources to produce food or fuel: Policy intervention or choice.* Paper prepared for the Fourth International Symposium on Alcohol Fuels Technology, Guarujá, SP, Brazil, Oct. 1980 (University of Sao Paulo, Instituto de Economia Agrícola).

Rathod, K.L. (1973). "Response of sugar-cane producers to prices: A case-study of western Uttar Pradesh," in *Agricultural Situation in India* (New Delhi: Directorate of Economics and Statistics), Sept.

Ray, Anandarup (1984). *Cost-benefit analysis: issues and methodologies* (Baltimore,

MD: The Johns Hopkins University Press).
Ray, S.K. (1976). *Variations in crop output* (New Delhi: Institute of Economic Growth; mimeographed).
______ (1981a). *Intensification of agriculture in the plains of Uttar Pradesh* (New Delhi: Institute of Economic Growth; mimeographed).
______ (1981b). *Changes in agriculture in the eastern and western regions of Uttar Pradesh* (New Delhi: Institute of Economic Growth; mimeographed).
Reddy, A. and K. Prasad (1979). "Technology alternatives and the Indian energy crisis," in *Economic and Political Weekly* (Bombay: Sameeksha Trust), Vol. XII, special Nos. 33 and 34.
Reynolds, J. (1981). *Management training cases in rural technology choice: low-lift irrigation pumps* (Geneva: ILO; mimeographed F/MAN/NOR Government Programme for Development of Managers working paper; restricted).
Rogers, P. (1981). *The economics of small producer gas-engine systems* (Cambridge, MA: Harvard University; mimeographed).
Roy, Rathindra N. (1982). "Integration, application, and management of energy activities at the community level: Technology transfer and communication," in R.C. May et al. (Eds.), *Appropriate technology for alternative energy sources in fisheries* (Manila: Asian Development Bank and the International Centre for Living Aquatic Resources Management; Conference Proceedings).
Roy, Sanjit (1983). "Six good reasons to go to the Tilonia way," in *Yojana* (New Delhi: Ministry of Information and Broadcasting), Vol. 27, n. 6.
Rural Electrification Corporation (1980). *Benefit-cost study of the RE scheme in Ghazipur* (New Delhi).
Ruttan, Vernon (1984). "Integrated rural development programmes: A historical perspective," in *World Development* (London: Pergamon Press), Vol. 12, n. 4.
Samantha, B.B., and A.K. Sundaram (1983). *Socio-economic impact of rural electrification in India* (Washington, DC: Resources for the Future; mimeographed; Jan.).
Satyanarayana, Y. (1967). "Factors affecting sugar-cane acreage in India," in *Indian Journal of Agricultural Economics* (Bombay: Indian Society of Agricultural Economics), Apr.–June.
SERA (1981). *Energy options and employment* (London: SERA; mimeographed).
Shannin, T. (1971). *Peasants and peasant societies* (Harmondsworth, UK: Penguin).
Shailaja, R., and N. Ravindranath (1984). *A field evaluation of a fuel efficient, smokeless woodstove: Astra ole* (Bangalore: Indian Institute of Science, ASTRA; mimeographed; restricted).
Silva, J. Gomes da (1976). "Balanço energético cultural da produçao de álcool etílico de cana de açúcar, mandioca e sorgo-sacarino: Fase agrícola e industrial," in *Brasil Açucareiro* (Rio de Janeiro: IAA), Vol. XLV, n. 6.
Singh, J.B. (1982). "Biogas programme: Involvement of voluntary organisations," in *Financing Agriculture* (Bombay: Agricultural Finance Corporation, Ltd.), Vol. XIV, n. 2–3.
Sinha, B., and R. Bhatia (1983). *Economic appraisal of irrigation projects in India* (New Delhi: Agricole Publishing Co.).
Sinha, S.P. (1982). *Apolly-12-PU-500 Windmill* (Bangalore: National Aeronautical Laboratory; mimeographed proceedings of Wind Energy Workshop).

Smith, D. (1980). "Rural electrification or village energisation," in *Interciencia* (Caracas), Mar.-Apr.

______ (1985). *Photovoltaics: socio-economic impact and policy implications* (Geneva: ILO; mimeographed World Employment working paper n. 151; restricted).

Smith, D., P. Hayes, and D. Mehta (1983). *Regional rural electrification survey.* Report to the Asian Development Bank, Bangkok.

Squire, L. and H. van der Tak (1975). *Economic analysis of projects* (Baltimore: The Johns Hopkins University Press).

Stewart, W. (1982). *Mud stoves* (Mt. Rainier, MD: VITA).

Stuckey, D.C. (1983). *Technology assessment study of biogas in developing countries,* paper prepared for World Bank/UNDP global project GLO.80603 (Duebendorf, Switzerland: International Center for Waste Disposal).

Subbarao, K. (1969). "Farm supply response—A case-study of Andhra Pradesh," in *Indian Journal of Agricultural Economics* (Bombay: Indian Society of Agricultural Economics), Jan.-Mar.

Subramaniam, S.K. (1977). *Biogas systems in Asia* (New Delhi: Manpower Development Institute).

Sugar-Cane (various issues). (New Delhi: Fertiliser Association of India).

Swedish International Development Agency (SIDA) (1982). *Report from International Seminar on Community Forestry* (Stockholm).

Swiss Development Co-operation Agency (SDCA) (1985). *Choice of Technology for lifting irrigation water* (New Delhi; mimeographed).

Takla, A. (1982). *Technological information profile on solar energy applications* (Vienna, United Nations Industrial and Development Organisation (UNIDO), Sep.).

Tata Energy Research Institute (1981). *Promotion of water-pumping windmills in developing countries* (Bombay: mimeographed; July).

Taylor, Robert P. (1981). *Rural energy development in China* (Washington, DC: Resources for the Future).

Taylor, L., E. Bacha, E. Cardoso, and F. Lysy (1980). *Models of growth and distribution for Brazil* (Oxford, UK: Oxford University Press for the World Bank).

Tewari S.K. (1981a). *An analysis of the appropriateness of wind and other energy systems of irrigation water pumping in India* (Amsterdam: Elsevier Scientific Publishing Company for the Agricultural Water Management).

______ (1981b). "A national programme for wind energy: Status and possibilities," in *Urja* (New Delhi: Dipak Chaudhuri, Editor and Publisher), July 19.

TOOL-ORP (1980). *Report on field evaluation of windmills in Ghazipur* (Ghazipur: TOOL Foundation), Dec.

Tyagi, D.S. (1974). *Farmers' response to agricultural prices in India* (Delhi: Heritage Publishers).

UNIDO (1978). *Guide to practical project appraisal: Social benefit cost analysis in developing countries* (New York/Vienna).

______ (1979). *Appropriate industrial technology for energy rural requirements* (New York/Vienna).

______ (1980). *Appropriate industrial technology for sugar* (New York/Vienna).

______ (1983). *Mini-hydropower stations* (New York/Vienna: Small Hydropower Series, n. 1).

United Nations (1980). *Technical Co-operation for Development (1984): Energy planning in developing countries* (Oxford: Oxford University Press).

United Nations Conference and Trade and Development (UNCTAD) (1983): "Handbook of international trade and development statistics" (New York: United Nations).

______ (1984). "Technology issues in the energy sector of developing countries: small-scale hydro-power projects in Nepal" (Geneva: report for the Committee on Transfer of Technology, Fifth Session, 22 October, TD/B/C.6/116).

United Nations Development Programme (UNDP)/World Bank (1981–85): *Issues and options in the energy sector,* Reports (by country) of the Joint UNDP/World Bank Energy Sector Assessment Programme (Washington, DC).

University of Allahabad (1980). *Cost and profitability aspects of sugar-cane cultivation in the Deoria District (UP)* (India: Agricultural Economics Research Centre).

Uttar Pradesh State Electricity Board (UPSEB) (1978). *Draft Annual Plan and Five-Year Plan for Power Development: 1978–83* (Lucknow).

______ (1980). *UPSEB statistics at a glance* (Lucknow).

______ (1981). *Achievements and problems of Uttar Pradesh State Electricity Board* (Lucknow).

van Buren, A. (1979). *Chinese biogas manual—popularizing technology in the countryside* (Ottowa: IRDC).

Vel, J.A. and L.R. van Veldhuizen (1981). *A model for the economics of small scale irrigation with windmills in Sri Lanka* (Amersfoort, the Netherlands: CWD; mimeographed; Oct.).

Vergara, M.W. and D. Pimentel (1978). *Fuels from biomass: Comparative study of the potential in five countries, advances in energy systems and technology* (London: Academic Press).

Walton, J., A. Roy, and S. Boman (1978). *A state-of-the-art survey of solar-powered irrigation pumps, solar cookers and woodburning stoves for use in sub-Sahara Africa.* Report to the al Dir'iyyah Institute (Atlanta, GA: Georgia Institute of Technology).

Watt, S. (1976). *Chinese chain and washer pumps* (London: Intermediate Technology Group).

Webb, S. (1981). "The impact of increased alcohol production of the agricultural sector: A simulation study," in *American Journal of Agricultural Economics* (Lexington, MA: American Agricultural Economics Association), Aug.

Wilson, S.S. (1977). "Pedal power on the land," in McCullagh, J.C. (Ed.), *Pedal power* (London: Rondale Press).

World Bank (1980a). *Energy in the developing countries* (Washington, DC).

______ (1980b). *Alcohol production from biomass in the developing countries* (Washington, DC).

______ (1981a). *Energy pricing in developing countries: A review of the literature* (Washington, DC).

______ (1981b). "Mobilising renewable energy technology in developing countries" (Washington, DC).

______ (1982). *Emerging energy and chemical applications of methanol: Opportunities for developing countries* (Washington, DC).

______ (1983). *The energy transition in developing countries* (Washington, DC).

Index

CPSIA information can be obtained at www.ICGtesting.com
Printed in the USA
LVOW10*1348121213

364936LV00006B/18/P

9 780275 928513